Praise for Earlier Editions

"*Power With Nature* demonstrates that a practical informative book on renewable energy doesn't have to be boring. It takes a technical subject and translates it into a how-to book with an added twist of humor. The entire book communicates well to both the layperson and energy enthusiast."
— *Mary Jane Masters, Institute of Ecolonomics*

"Readers who want to be completely self-reliant, who are looking for cost-effective solutions, or who are somewhere in the middle—all will find something useful here...Ewing writes in a folksy, informal manner, and readers will find his hands-on primer worthwhile."
— *Library Journal*

"Plenty of energy guides for homeowners advocate getting off the grid; but few do such a good job of explaining just how to go about it... An outstanding, highly recommended guide."
— *California Bookwatch*

"Should be mandatory reading for all environmentally conscious folk. This book is immensely useful and readable, and enjoyable. Whoever says conservation and science are stuffy subjects needs to have this book to reverse their view."
— *amazon.com reviewer*

"It is a fun book, easy to read, delightful illustrations; the appendices are worth the price of the book themselves. And I appreciate the bonuses, e.g. calculating rainwater collection and info on corn and pellet stoves. Congratulations for producing a really great handbook for renewable energy enthusiasts!"
— *H. Dana Moran, Colorado Energy Science Center at NREL*

"I've read virtually every book that has been written about using solar and wind to power your home and this book is one of the best. I gave it to my wife, a true solar neophyte, and here are her comments: 'I hate technical manuals, but this book is easy to comprehend from the perspective of someone who knows next to nothing about electricity and how it works. I would highly recommend it to anyone interested in building a self-sufficient home or changing to a grid-tie system.' "
— *Larry Cooper, Kyocera Solar, Inc.*

"Your solar book is so timely. It not only explained how solar systems work for a dwelling, but proved to be a good reference point for the immense topic of electrical systems in general. I now have a good working knowledge of these things and feel confident I will no longer have the deer-in-the-headlights look when listening to a salesman before buying."
— *Barb Fabisch*

OTHER NON-FICTION BOOKS BY REX A. EWING

Crafting Log Homes Solar Style

Got Sun? Go Solar

HYDROGEN—Hot Stuff, Cool Science

Logs, Wind and Sun

Beyond the Hay Days

Power With Nature

Renewable Energy Options
for Homeowners

UPDATED 3RD EDITION

REX A. EWING

PIXYJACK PRESS INC

Power With Nature: Renewable Energy Options for Homeowners
Updated 3rd Edition

Copyright © 2012 by Rex A. Ewing

Published by PixyJack Press, Inc.
PO Box 149, Masonville, CO 80541 USA

3rd Edition © 2012
2nd Edition © 2006
1st Edition © 2003

ISBN print edition: 987-0-9773724-9-2
ebook also available

Library of Congress Cataloging-in-Publication Data
Ewing, Rex A.
  Power with nature : renewable energy options for homeowners / by Rex A. Ewing. -- Updated 3rd edition.
      pages cm
  Includes bibliographical references and index.
  Summary: "Covers renewable energy options for grid-tied and off-grid home-owners, including solar energy (passive and active), wind power, microhydro energy, geothermal heat pumps, solar water heating, biomass heating, backup generators and pumping/storing water. Also examines energy conservation, system sizing/pricing, and tax incentives"-- Provided by publisher.
      ISBN 978-0-9773724-9-2
  1. Solar energy. 2. Wind power. 3. Renewable energy sources. 4. Distributed generation of electric power. I. Title.
      TJ809.4.E95 2013
      696--dc23
                      2012042277

For LaVonne Ann—
you fill my life with magic

Contents

continued

Appendix

THIRD EDITION

Prologue

I have always thought renewable energy an adventurous enterprise. It certainly was back in 1999 when LaVonne and I moved to the mountains and cobbled together our first solar-electric system. Solar and wind installers were few and far between, and if you did happen upon one it was a fair bet that he or she had a day job and was doing renewable energy installations on the side. This was a time when Trace Engineering was the only name in power inverters and charge controllers and no one complained about it; when electrical utilities did not know what to make of solar power but they were all pretty sure they didn't want it contaminating their power lines; when the words "global" and "warming" were just beginning to be spoken one after the other and it was still possible to pick up a newspaper or a science journal without someone remarking about your personal complicity in bringing about the end of the world.

But things have changed. Today you can hardly walk down the street without bumping into a solar installer with certifications from one or more quasi-government institutions. Trace is now Xantrex and they're up to their eyeballs in competition. Practically every utility in the nation talks up solar and wind energy like they invented the stuff. And years of political and scientific horse whipping has cowed us all into believing that the world would be much better off if none of us had ever been born.

This is not all bad, of course. Competition in the solar marketplace has led to the creation of some truly wondrous new components with capabilities few could have anticipated a decade ago. Photovoltaic modules are now cheaper and better than ever, and they can be connected to one another by anyone who can shove a round peg into a round hole. Yet with such a wealth of components to work with, solar and wind systems have necessarily become more complex, hence the emergence of the trained and certified solar installer— that eminently qualified person you would want to install your system in the event you decided not to do it yourself. And, of course, with so much solar and wind energy out there we should all be grateful there are utilities ready and willing to buy it from Peter and sell it to Pauline.

Yet somewhere along the line, an idea that was once fun and adventurous has begun to resemble a political and "moral" mandate, and now solar and wind technologies are burdened with the task of saving the world from the perils of human meddling. It's the sort of thing that happens when scientists crawl into bed with politicians.

Power with Nature is not about political agendas or scientific prophesies, however, and nowhere will I denigrate you for your addiction to gasoline, your love of red meat from methane-farting cows, or the embarrassing size of your carbon footprint. It's not that kind of book. For while I am indeed proud to be personally involved with the clean technologies that will one day replace fossil fuels, that's not the reason I'm in it. It's much simpler than that: I like being in control of my own energy usage and production, and I thoroughly enjoy dreaming up energy-conserving strategies and tactics that would never occur to anyone not living off the grid with limited energy resources.

I therefore hope to show you in some detail the most promising renewable energy technologies and how they can save you money and make you more self-reliant. I would also like to instill in you the same sense of wonder I feel every time I wire together a solar array or a battery bank, or climb my wind tower for a little routine maintenance on my Bergey XL-1 wind turbine. Because for all the things renewable energy may or may not be, it is, first and foremost, really fascinating technology that has the added benefit of being a helluva lot of fun. If I am able to get that much across, then I will have done my job.

The first two editions of this book began with a fable titled *Dog of the Sun, Cat of the Wind*. It was about my improbably talented cats and dogs, my search for the perfect woman, and, naturally, the bare rudiments of renewable energy.

But like most works of fiction, it has become dated and, I daresay, politically incorrect in that I shamelessly promote off-grid over grid-tied systems. And while I certainly have no problem with political incorrectness—indeed, I practice it every chance I get—the fable was in need of some technical revisions and I didn't have the heart to change a word of such a finely spun tale. It is not dead, however; it is waiting in PDF format as a free download at *www.PixyJackPress.com* for all who would venture to read it. If nothing else, it will shed a great deal of light on the many morsels of dog and cat wisdom sprinkled throughout the book. *Enjoy.*

FABLE ILLUSTRATIONS BY SARA TUTTLE

...The scene before us as we pulled into the yard was like something out of a John Carpenter film. Packrats were pouring out of the hay barn and scrap piles, the equipment shed and wood piles in droves, their little black beady eyes glistening in the headlights as they scurried toward the house. The three dogs and Stinky, the cat, had tried to set up a defensive perimeter around the house, but there were too many rats for the four of them to fend off; whenever one of them would send a rat running, another five would sneak past.

Through the window I could see that Willie had managed to push my office door shut, ensuring that no dog could deflect him from his twisted mission. He was sitting on his haunches on my desk, flipping the desk lamp's light switch on and off with his paws, sending out his diabolical message to all his little nether-minions. His face was frozen into a maniacal rictus.

...If Willie was surprised to see me, he didn't let it show. The leer I'd seen through the window was still hard-set on his face, as though cemented in place by Igor, the hunchback taxidermist. However, the second I screamed, "Prepare to die, you good-for-nothing cat!" and lurched for his scrawny neck with outstretched fingers-become-claws, he realized the party was over. Like a furry ball of yellow lightning, he leapt from the desk—scattering hundreds of pages of the novel I'd been working on—zipped out of the office, and ran through the wide-open front door. Though it was obvious—even to me, in my advanced state of rage—that I would never catch him, I took off in hot pursuit...

To read the entire fable, visit www.PixyJackPress.com

Introduction

A New Philosophy of Freedom
Learning to Think Beyond the Grid

Sometimes life's changes unfold slowly, one year rolling into the next almost imperceptibly as time works its furtive magic. Other times, the changes come quickly and willfully, as in 1999 when LaVonne and I sold our horse ranch on the Colorado plains and moved into a small mountain cabin, with plans of building a log home as a permanent residence.

Upon moving we quickly realized we were facing a sudden and irrevocable change in lifestyle. The fact that we had just sold a 2,000 square-foot house so we could live in an amenity-free cabin with the footprint of a one-car garage was the least of it. The real challenge was learning to deal with the sum of all the other changes.

Where before we had a well that pumped water at the rate of 30 gallons per minute, we now had to drive 20 miles to town once a week to fill a 200-gallon tank roped into the back of a pickup. Our toilet was an outhouse, our bathtub a creek. A large cooler had replaced our spacious refrigerator (sorry, no more ice maker). We now used a wood stove for heat, a gasoline camp stove for cooking, and kerosene lanterns for light.

And when we simply *had* to have electricity, I would go solemnly into battle with a war-hardened, 4,000-watt Coleman generator. If I prevailed over the surly beast, then we could run a saw or a vacuum. If I was bested, then we just had to wait until I could sneak up on the smelly thing and yank its cord before it had a chance to suck in a carburetor-full of gas, belch flames, and flood itself.

One would think that such an abrupt "lowering" of living standards would manifest itself as individual stress or even marital strife, but nothing of the sort occurred. Quite the opposite, in fact; more than anything, LaVonne and I embraced our new life with all the energy and ambition of a couple of kids on an extended camping trip. Having purged our lives of the leaden inertia that crystallizes in the consciousness of anyone in the habit of having an instant remedy for any earthly desire, we quickly came to appreciate every drop of water, every morsel of food, every ray of light in the midst of darkness.

The First Precious Amenities

We were, quite literally, starting over from scratch. Any comforts or conveniences we hoped to enjoy would have to come as a direct result of hard work and mutual cooperation, not from wringing our hands over our plight.

Living just a notch or two above primitive, we quickly prioritized our desires. (I say "desires" because we actually didn't *need* anything more than we had.) First on our wish list was hot water. And not the kind you heat on the stove in a brass pot for a cup of tea; we wanted hot water flowing copiously from a showerhead, inside a closed shower stall, within a warm room, walled off from the clouds and the wind.

Toward that end, we built a small 6- x 16-foot addition on the back of the cabin; just big enough to hold a tiny shower stall, a used 35-gallon propane hot water heater, a propane refrigerator (which, as it turned out, wouldn't arrive for several months, thanks to the Y2K mania that was sweeping America at that time), three water barrels, and everything we needed for a simple solar-electric system.

We pressurized the fresh water system with a small 12-volt RV pump, which was powered by a pair of 12-volt deep-cycle batteries; one hooked to the pump, the other was wired to a small solar module for charging. When the battery in use ran down, we switched them out. (We have since installed a DC-to-DC converter to run the 12-volt pump from the main 24-volt battery bank, as our stand-alone deep-cycle batteries have at last worn out.)

The electrical system followed the plumbing and gas piping. Using the solar modules and power inverter we'd purchased for the log house—which in late spring of 1999 was no more than a muddy hole in the ground—we had more power than we ever could've used in that small cabin, even during a run of cloudy days of biblical proportions.

Considering that until then my most meritorious achievement as an electrician was running AC power to a few stock tank heaters for my horses, the complexity of that first bare-bones PV system seemed daunting. I read every manual front to back, one, two, three times. I put every component of the system into place, then slowly and meticulously hooked-up my wires, using a multimeter to check and double check every step along the way, testing every connection ten times over before finally flipping the switch to the DC disconnect and turning on the inverter. No sparks, no explosions. Just a steady hum and the soft, green glow of the display. After testing the AC side of the system—to see if my multimeter was as convinced as I was that we actually *were* producing usable power from the sun—I began trying different loads. First a small light, then an electric drill.

Impressed, but still not convinced, I opted for the ultimate test: a voracious 15-amp table saw. It was the single piece of equipment that could make the burly old Coleman generator convulse with fear (of a herniated head gasket, I would imagine). I plugged it in, hit the switch (after a fleeting, jumbled moment of pensive hesitation) and watched with consummate awe as the blade spun quickly and effortlessly into motion. Though my left brain knew the whirring saw blade was merely the logical outcome of applied technology, my right brain insisted I had just witnessed a miracle.

Regardless of which interpretation one chooses to embrace, it was in that instant when the table saw came to life that the idea of sustainable, free power from the sun was transported from the theoretical realm to the practical.

Moving Up

By the time we were far enough along on the new log house to run wires and install the photovoltaic (PV)/wind system, I thought I knew practically everything there was to know about solar energy. It was a thoroughly absurd notion, of course; kind of like the teenager whose first distant glimpses of adulthood lead him to overlook life's myriad subtleties and draw the erroneous conclusion that life is a simple subject and any grownup who doesn't agree is an idiot.

The first cracks in my thin, hard shell of ignorance came when our electrician—after not showing up for several weeks—made the grievous mistake of mouthing-off to LaVonne.

He'd have been better off poking a wildcat with a sharp stick.

With the building-boom in town, there was zero chance of finding another electrician who would be willing to drive up into the hills anytime soon. And we couldn't wait. So, (very) reluctantly, I told my wife that, given enough time—and enough books on the subject—I could finish wiring the house and installing the solar and wind systems. And, though I nearly choked on this part, I told her I could do it all to code. I didn't have much idea just what the National Electric Code was in those days, but I was pretty sure I was about to find out.

On my first inspection after countless hours of work, the inspector wrote me up for 18 violations. He later called back and admitted he was wrong about two of them, leaving me with only 16 violations to deal with.

In an attempt to ensure that my next inspection wasn't an encore performance, I must have ended up talking to every high-level electrical inspector in the state at one time or another. No one remembered my name, but they all knew "the guy with the wind turbine," since it was one of the sticking points in our negotiations. The main problem was that code required all sources of electrical power to have a manual disconnect, while design requires a wind generator to be connected to a load at all times. In the end, design won out over code.

Happily, my wiring passed the next inspection without a hitch; for the first time in two years, we would soon be living in an electrically correct house. It was only then that my real education began.

Learning the System

Once we moved into the new house we realized that even after doubling our generating and storage capacities we now possessed the means of using energy faster than we could produce it. The well pump was the main culprit, followed by the dishwasher and the hot water circulating pumps for the propane-fired radiant heat boiler. (The usual suspects, namely the refrigerator, range and clothes dryer, were all powered by propane.)

It quickly became clear that a means to monitor our energy usage was necessary—for peace of mind, if nothing else. We checked around to see what was available, then bought a TriMetric Meter from Bogart Engineering. It was a little tricky to install and calibrate, but well worth the trouble. Not only can we use it to see roughly how much wattage each of our appliances is consuming, it also keeps track of amp hours going into the batteries versus amp hours going out, providing a digital fuel gauge for the system. (The TriMetric performs a lot of other useful functions, as well. It gets more attention than any other component in the system.)

Once the batteries are full, the charge controllers for both the wind and solar systems back off the power delivered to the batteries. It is, therefore, most efficient to use all the energy you can while it's there for the taking. Make hay (wash clothes, run the dishwasher) while the sun shines, as the expression goes.

Seeing how the wind plugs into our energy equation has been as fascinating as it has been instructive. While the solar array is the real workhorse of the system, the wind is like a whimsical sprite that often shows up just when we need it most. Many people have remarked that extra solar modules would have been cheaper, delivered watt for delivered watt, than our wind generator and tower. And they're right. But they also completely miss the point, since the wind most often provides power at night and during stretches of cloudy weather, when the solar array is idle, or nearly so. This means that we can get by with less storage capacity than any of our solar-only neighbors, since we're charging our batteries while they're depleting theirs—a fact that is particularly gratifying after three days of cloudy, windy weather.

Daily I feel more respect for our wind and solar system and its remarkable ability to rejuvenate itself. On most days our batteries are fully charged before lunch, even while running two computers and a stereo all day long, plus any other tools or appliances LaVonne and I can throw into the mix.

Consequently, my attitude toward energy usage has become much more relaxed, since I know that we'll gain it all back in due course. LaVonne, too, has taken notice of my moderated vigilance over the wattage reserves. She hardly ever calls me an Energy Nazi anymore.

It's just a matter of learning to trust Mother Nature.

– / –

Overview of Options and Incentives

High- and Low-Tech Ways to Becoming Self-Sufficient
and Getting Paid For It

Whether your goal is to trim the rough edges off your carbon footprint, free yourself from the clutches of the coal and nuclear industries, or simply to save a lot of money, renewable energy is the only practical way to go in a world where energy is getting scarcer by the day. Sunlight, wind, flowing water, and heat from the Earth are all free, as is biomass (in the form of cordwood) if you happen to have access to a nearby forest. But there's a lot more to it than that.

To make renewable energy truly effective you also have to conserve energy wherever you can. There are a multitude of ways you can do this, such as sealing leaky ducts, doors and windows, adding insulation, and installing programmable thermostats. Get an energy audit of your home. These clever folks will find energy leaking from places you didn't even know existed. Conservation begins with being aware of your energy usage and taking steps to put you in charge of making you and your family more self-reliant.

These steps go hand in hand with the more high-tech solutions of adding a solar-electric array, which you can own outright or, in some states, lease. Or installing solar thermal collectors to heat your domestic hot water. Or putting in place a super-efficient geothermal heating/cooling system. The fact is, you now have more choices than ever. And those choices are getting more affordable and more efficient every year.

In this updated edition, we'll explore a wide variety of options for all budgets, from passive solar and other smart building techniques, to generating electricity with solar, wind and microhydro systems for grid-tied or off-the-grid homes. You'll discover new options for generating heat and new ways to efficiently pump and store water. And in between you'll learn a bit about the equipment and safety features employed by all the different technologies.

If you live out in the hinterland you have more options—and more specific needs—than city folks, especially if you've chosen to live off-grid. As you just read, we've been trusting Mother Nature to provide since 1999. Thus far She's done a bang-up job for us, and She's taught us a lot of valuable lessons along the way, not the least of which is the agreeable fact that getting by on less can be far more rewarding than always having more than enough.

But personal philosophies aside, renewable energy just makes good sense; now more than ever. Today there are far more ways to make conservation and renewable energy pay than there were 10 years ago. For not only will your efforts enhance your property values, they can also earn you some very tangible benefits in the form of tax credits and rebates just for doing the things you should be doing anyway. So, before we plunge into the vast sea of renewable energy technologies, we should take a little time to explore the paybacks.

Our ground-mounted solar array south of the house and the wind turbine to the east.

Return on Your Investment

Installing a solar electric system increases the value of your home, particularly as Americans continue to become more energy conscious. According to a now somewhat dated but extensively researched study from the National Appraisal Institute (*Appraisal Journal*, Oct. 1999), your home's value increases $20 for every $1 reduction in annual utility bills. This is a ratio that will only increase as energy costs rise. At 1999 rates, a modest solar-electric system of 2,500 watts will increase your home's value by $8,000–$10,000 immediately, just for the utility bill reduction. Many states have adopted property tax exemption laws, such that if your home value is increased by a solar installation, you cannot be taxed on that increased value. You'll find details at the DSIRE website *(www.dsireusa.org)*.

Financial Help

The good news is that you will not have to bear the entire cost of renewable energy systems. There are utilities and government entities out there that want to give you money for your foresight and sensitivity to environmental issues. My advice is to let them.

Federal Incentives

First, the incentives that apply to all Americans. Beginning with *The Energy Policy act of 2005* and culminating with *The American Recovery and Reinvestment Act of 2009*, the federal government has outlined in fine detail just what it feels it is worth to the country for you to invest in renewable-energy technologies for your home. As it turns out, it has concluded that 30 percent is fair compensation and it is offering it in the form of an income-tax credit until December 31, 2016.

What do you have to do to get it? Since the appeal of reading unabridged legislation ranks right up there with DIY dental work for most people, I suggest you visit the Database of State Incentives for Renewables & Efficiency at *www.dsireusa.org*. This website is meticulously maintained and updated by the folks at the North Carolina Solar Center, which is operated by NC State University. They have translated all the legalese into terms the rest of us can understand and have organized the information in impeccably logical order, making it quite simple to determine who is willing to incentivize you to invest in renewable energy.

As you will see, the 30% federal tax credit applies to a broad range of technologies with surprisingly few restrictions. Among the technologies discussed in this book, the tax credit extends to 1) solar-electric properties; 2) solar water-heating properties; 3) small wind-energy properties; and 4) geothermal heat pumps.

Each of these installations will allow you to deduct 30% of the installed cost of the system from your income taxes for the year the system was installed, with no upper limit for systems installed after 2008.

For solar hot-water systems the equipment must be certified by the Solar Rating Certification Corporation (SRCC), and it must provide at least half of the hot-water used by the residence. Nor, for reasons that should be obvious, can a credit be taken for systems designed to heat swimming pools or hot tubs.

For geothermal heating/cooling systems, the heat pump must meet federal Energy Star criteria.

Biomass-burning stoves and fireplace inserts may qualify for a residential energy-efficient property tax credit of up to $300.

State and Local Incentives

State and local incentives are also listed on the DSIRE website (*www.dsireusa. org*). Generally these come in the form of rebates from state and local governments, and from utilities scrambling to buttress their renewable-energy portfolios. Bear in mind that there is only so much money available for each program and once it's gone, it's gone. And, as a rule, filling out the paperwork can be a lengthy, hair-pulling process, especially if you've never done it before.

Fortunately for those of you who long ago concluded that you would be better off hiring someone else to do the work, a professional installer will not only know exactly which programs are currently active in your area, he or she will also fill in the forms. And many will even accept the promise of money due as a down payment on the work. It certainly simplifies things.

PACE Programs

Another option that is becoming more popular for property owners who want to get into solar energy for little or no initial investment is Property Assessed Clean Energy (PACE) financing. Under these programs the city or county government pays for your system upfront and is reimbursed over a specified term through a surcharge on your property taxes. You own the system, however, and will be able to reap the full benefit of the electricity it produces. You will also receive any and all rebates and tax credits due to you from local, state and federal programs.

PACE programs are not limited to solar-electric installations. They are usually structured to include a wide array of renewable and energy-saving technologies, from wind turbines, solar hot water and geothermal, to daylighting, efficient window and doors, insulation...the list goes on. To see if any PACE programs are available where you live, visit the *www.dsireusa.org* website.

. . . .

Got Insurance?

You'll want to tell your homeowner-insurance carrier about your renewable-energy equipment—just to make sure everything is covered. Our policy, for instance, covers anything mounted on the roof as part of the structure, but the ground-mounted PV array is protected under the 'dwelling extension' coverage. Don't forget to mention your batteries, inverter and other components, wherever they may be located. Know what your risks are. Living on a ridgetop, we worry about lightning strikes, damaging winds, wildfire...and the neighbor's cows. You may have additional concerns where you live.

. . . .

MICK'S MUSINGS

Watching Rex install the solar and wind systems was instructive. I learned a whole bunch of new words.

~ 2 ~

Building Smart to Keep the Power Bill Low
A Little Planning Can Reap Big Rewards

Energy is slippery stuff and that makes it a problem. Your house loses heat at night for the same reason an ice cream cone melts all over your hands and drips into your lap: everything wants to become the temperature of the surrounding environment. What was considered adequate building insulation in the days of cheap energy is wholly unsatisfactory today. Insulate, insulate, insulate! And while we are told that sealing tightly around doors and windows and ductwork makes good sense, it really does pay off in energy savings. Designing and building a house today without carefully considering how best to make it energy efficient is like deciding you'd like to make charitable contributions to your local utility company. Every month. Forever.

For those planning to build a home, every dollar you spend on good home design will save you twice that on renewable energy equipment. So the old proverbial drawing board is the place to start if you hope to make the most of the energy you use.

Orientation of Home and Floor Plan Considerations

Unless you plan to build a dome house, or some other design that deviates from conventional building practices, your house will have a long axis and a short one. For the purpose of utilizing passive solar energy—and to make things easier on yourself, once you get around to installing active solar PV modules and collectors—you will want to run the long axis of the house east and west to maximize the amount of solar radiation that enters the house in the winter months.

If you are a few degrees off it really won't diminish the efficacy of passive or active solar systems. So if you know the magnetic declination for your area (since magnetic north and true north differ from one another by varying and changeable degrees in different places) you can use a compass to get a pretty close approximation of the cardinal points. On the other hand, if you are truly persnickety—I know I am—about the alignment of things, you can determine the true north-south axis with a couple of T-posts and a clear night sky *(see 'Where is North?' in chapter 5)*.

With the long side of your house facing south, the area exposed to direct solar radiation in winter will be greatly increased. There are a couple of things you can do to make the best of this free energy. An open floor plan will allow solar radiation greater access to the deep recesses of your house.

Thermal Mass and Trombe Walls

Of course, to really take advantage of the sun's low angle in winter, it's not enough just to point your house's broadside in the sun's direction and hope for the best; you have to invite it inside where it can really work some magic. This is because light coming in through a home's windows is absorbed by things inside the house and reemitted as heat (which is simply light with a longer wavelength) that cannot easily pass back through the glass. This is the principle behind the much-publicized greenhouse effect. To ensure that this free heat is not squandered on furniture, carpets and dogs (none of which is all that good at storing heat), you want to place things in the path of the light that have appreciable thermal mass. What sorts of things? Well, the most obvious is ceramic floor tile. Tile can soak up a surprising amount of heat, and if it's laid over regular or gypsum concrete, as it would be for a highly efficient hydronic in-floor heating system (hint, hint), the heat gain is further enhanced.

A Trombe wall is also a terrific heat-storing design feature. Consisting in principle of nothing more than a concrete wall, or wall segment, installed within an inch or two of a south-facing window and sealed around the edges, Trombe walls are showing up in more and more homes these days, and for good reason: it takes the heat trapped between the wall and the glass 8 to 10 hours to pass through an 8-inch thick concrete Trombe wall into the interior of a building, so you'll just start feeling the heat of the noonday sun a little before bedtime. It's the stuff sweet dreams are made of.

Other effective indoor heat reservoirs include rock and/or concrete planters, or even large water-filled metal columns coated with unreflective paint. You can let your imagination go wild on this one.

Natural Daylighting

If you are building a new home, you'll have plenty of options for optimizing passive solar energy long before you hook up your active solar array. LaVonne and I looked at a lot of houses before designing ours. Many of the older log homes we saw back in the woods were little more than fortresses with tight, economic floor plans, low-pitched roofs, and conspicuously few windows, all barely big enough to aim a rifle through. Many seasoned homes in town were not much better. They were built back in the days when windows were considered to be a necessary source of heat loss. These old homes were well designed to conserve any heat produced within the walls—the dearth of insulation and multi-glazed windows notwithstanding—but woefully deficient at allowing heat in from the outside.

The advent of efficient double- and triple-glazed windows has changed that myopic view of home design. Realizing the potential of free solar heat in the winter, most homes today—outside of those built in cheesy developments—take ample advantage of the sun's gifts. Prow-like projections with giant windows set between massive frames adorn the southern faces of many modern homes.

This is all well and good, but like everything else, it's possible to get carried away. An overabundance of glass will certainly keep a house warm on sunny winter days, but it will also allow excess heat leakage at night or when the weather turns cloudy. If you are building a new home, you should strive for a balance. Design your house to allow ample sunshine on the south side, with as few doors and windows as possible on the north. Large windows, glass doors, and dormer (or gable end) windows that follow the contour of the roof all allow in plenty of sunshine while enhancing the home's appearance.

East- and west-side windows will each be in the sun for half of the day, but there is a difference. Whereas the east side will experience the cooler morning sun during hot summer months, the west side will be mercilessly baked when the afternoon sun kicks into overdrive. We know through experience that the finish on our log house takes considerably more beating on the west side compared to the east, and in similar fashion the west windows are the biggest source of unwanted heat on summer afternoons. Wherever you live, you should keep this disparity between east and west in mind when considering window placement in a new home.

If winter is a long, cold ordeal where you are planning to build, you might want to consider using triple-glazed or super-insulated windows. They allow as much light to enter as conventional double-glazed windows, but will retain appreciably more heat when the sun sets. Insulated window coverings are also extremely helpful for nighttime heat retention.

For those living off-the-grid, ample natural light is a must. A well-designed house should never need daytime lighting except on the cloudiest days. For interior rooms with no windows, light tubes provide a remarkable amount of light, even when it's heavily overcast. We recently installed a 10-inch light tube in a large storage shed. Along with one small window on the north side, the light tube provides all the light we need. Skylights can accomplish the same effect, although they do take up more roof space and tend to allow a lot of unwanted heat to enter a room. LaVonne had me put a small skylight in the loft walk-in closet. She very rarely needs to turn on the light as she fishes around looking for just the right ensemble on the rare days when we venture into town.

A sunny south side provides plenty of natural daylight and warmth on cold winter days.

Window Types: Climate Specific Options

What should you look for when selecting windows for your new home? There are a number of factors that affect window performance: the material used to make the frames; the number of glazing layers comprising the window; whether or not the in-between spaces are filled with a noble gas, such as argon or krypton; and which types of light-selective coatings will work best in your particular neck of the woods.

Frame types are pretty straight forward: because of its well-known conductive properties, aluminum is at the bottom of the list, while vinyl and fiberglass sit comfortably at the top. But don't despair; in terms of efficiency, wood and wood-clad windows (the kind you really had your heart set on) run a commendably close second.

The number of glazing layers is a no-brainer: the more the better, no matter where you live. Typical top-of-the-line windows have three.

Nor does it take much thought to determine the type of gas you want occupying the spaces between the panes. Ordinary air is made up of lightweight molecules of nitrogen and oxygen, both of which are extremely rambunctious compared with the draggletailed gases argon and krypton. What's the difference? Simply that heat-carrying convective currents are far more prevalent and robust inside air-filled windows, and it is because of these currents that long-wavelength infrared light (otherwise known as heat) is able to move from one side of a window to the other. By replacing frisky air molecules with a more somnolent gas, the convective conveyer is all but shut down.

Shorter-wavelength visible light, by contrast, easily passes through a medium of argon or krypton or, for that matter, empty space. Once inside the room it is absorbed by walls, floors, furniture, people and pussy cats, and re-emitted as long-wavelength heat. This is the general principle behind the greenhouse effect, the main difference being that when a planet becomes the room, its atmospheric layer becomes the windowpane.

A window's thermal properties can also be tempered by applying certain types of coatings to the glass. The effects of these coatings, combined with the sum of the above-mentioned factors, are measured by the Solar Heat Gain Coefficient (SHGC) and the window's emissivity, or U-factor.

Both the SHGC and the U-factor are determinations of the flow of energy beyond the boundaries of visible light, but at different ends of the visible spectrum. A lower U-factor is desirable, and for exactly the same reason we want our windows filled with sluggish gases. Low-Emissivity (Low-E) window coatings help to lower a window's U-factor by blocking the passage of heat through it, thereby reinforcing the insulating effects of argon or krypton.

By now you may have noticed something interesting. Every consideration thus far has related to a window's ability to discourage the passage of heat which, as I already mentioned, is a type of light that cannot be seen because its wavelength is too long for the human eye to discern.

Coatings that lower the SHGC however, block incoming sunlight in the short-wavelength ultraviolet range. Since UV light is every bit as invisible as heat (infrared), you can block as little or as much of it as you please without affecting the amount of illumination a room receives. This is fortunate, because once inside a room UV light is absorbed and reradiated as infrared. So by controlling the amount of UV light that enters a room, you can control how much or how little that room will be warmed by the sun.

Finally, we have a variable that can by adjusted to fit a particular climate. In cold regions high-SHGC windows are desirable because they let in as much heat as possible, whereas low-SHGC windows are more practical in warm climates where sunlight is appreciated but the heat resulting from it is not. Climate should not be the only consideration, however. Even warm climates can have chilly winters and high-SHGC windows will let in welcome heat through south-facing windows during the cold part of the year. The trick is design the home's eaves in such way that the harsh summer sun is forbidden direct entry through those same windows.

So which windows are the best match for your new home? To find out visit the Efficient Windows Collaborative at *www.efficientwindows.org*. Their user-friendly Window Selection Tool will instantly provide you with efficiency data for 30 or more different window types for any U.S. city you choose. You will quickly see the disparity between good windows and mediocre ones in terms of how much extra you will spend to heat and cool your house by choosing one type of window over another.

WILLIE'S WARPED WITTICISMS

Soft couches and south-facing windows have made cats nature's most effective solar collectors.

What you won't see is how much you are helping the planet by making the right—and ultimately, the economical—choice, but this may help: every dollar not spent on electricity saves the production of 20 pounds of carbon dioxide, and every gallon of propane not burned prevents the release of another 12 pounds. And all the while you're not producing CO_2 you'll be basking in the comfort your new windows will provide.

....

Help for Older Windows

If your windows seem to be a bit drafty around the edges, check into replacing the jamb liners. Our friends found it made a huge difference in the heating and cooling of their 20-year-old log home. They also installed Madico window film on their south and west windows; this film blocks up to 86% of the sun's heat and up to 99% of harmful ultraviolet rays that discolor rugs, floors and furniture. A small investment that brought big rewards year-round.

....

Big Eaves

Big eaves are a practical addition to a new home, since they keep direct sunlight from hitting the south-side windows in the warmer months. Then, in the winter when the sun drops low on the horizon, an abundance of solar radiation brings welcome warmth on cold, sunny days. Another bonus of big eaves: they will protect the sides of your house from rain and snow. With a little work, you can plot how many degrees above the horizon the sun will be at midday during different times of the year. Then, using basic geometry, you can easily calculate how long your eaves will need to be to keep your southern windows in the shade from late spring to early fall.

Natural Cooling

What about cooling? Air conditioners use about 5% of all the electricity produced in the U.S., at a cost of over $11 billion to homeowners. Air conditioning is an energy-intensive technology and if you are planning to live off the grid I urge you to look at other options, such as evaporative coolers, which do not need energy-hungry compressors. In low-humidity climates evaporative coolers use a quarter to one third the energy required by conventional AC units, but as relative humidity climbs above 50% their performance drops precipitously. For grid-tied folks, air-conditioning will make a considerable dent in your energy usage, but the new high-efficiency units are a big improvement over units from just 10 years ago.

The same windows that heat your home in the winter can help cool it in the summer (providing, of course, the house has the above-mentioned protective eaves). The key is to provide adequate cross-ventilation, preferably of a type that allows air to enter near the floor and exit through the roof. Double-hung windows, or windows designed with built-in vents at the bottom are perfect for letting air into the house. To provide an exit for hot, rising air, skylights—the ones that open—can't be beat; they are perfect for maximizing airflow through the house. Skylights also help to bring additional, natural light into a house, something that is often needed in lofts with few windows. Just don't install too many of them or you'll defeat their purpose. Skylights are great for letting in summer sunlight and letting out hard-won winter heat. Fortunately, this problem can be minimized by purchasing skylights with blinds. Conversely, many homeowners install north-facing skylights; they allow ample light while preventing the infusion of too much heat from the scorching summer sun.

Insulate & Seal

As we all know, insulation keeps your home warm in the winter and cool in the summer (if you're not convinced, spend a year in a drafty house trailer with 3-inch walls). Insulation works best when air is not moving through or around it, so to get your money's worth be sure to seal air leaks before installing insulation. Sealing air leaks around doors and windows is easy enough, but a bit trickier in attics, basements and crawl spaces. Common air leaks into the home include the dryer vent, kitchen stove vent, outdoor faucet, sill plates, and piping for air conditioning. Leaks also occur around bathroom vents, plumbing vent stacks, recessed lights, attic hatches, and duct registers.

In a typical house with forced-air heating and cooling, as much as 20% of the air moving through the duct system can be lost due to leaks and poorly sealed connections. Some ducts are concealed in walls and between floors and repairing them is well-nigh impossible. But exposed ducts in attics, basements, crawlspaces, and garages can be easily repaired with duct sealant. Insulating ductwork that run through spaces exposed to temperature extremes can likewise save a lot of energy. Then, just to be on the safe side, have an HVAC technician make sure that your combustion appliances (gas- or oil-fired furnace, water heater, and dryer) are venting properly.

How much can this save you? The EPA's *www.energystar.gov* website states that a knowledgeable homeowner or skilled contractor can save up to 20% on heating and cooling costs (or up to 10% on the total annual energy bill) just by sealing and insulating. My father-in-law agrees: after he built a tightly-sealed, super-insulated workshop, a string of 100-plus-degree days could not raise the temperature in his workshop above 73 degrees. And during long North Dakota winters, it remains quite cozy, with very little additional heat.

....

Enertia Homes: An Envelope of Glass and Wood

Most rectangular homes are enclosed by four primary walls. Some, however, such as those manufactured by Enertia Building Systems *(www.enertia.com)*, have six. Built entirely of laminates of sustainable Southern yellow pine—a dense, resinous wood noted for its heat-storing properties, Michael Sykes' designs synergistically combine the effects of several physical principles, including the greenhouse effect, the thermal inertia (mass) of wood and earth, and the natural convective movement of heated air.

So why the extra two walls? They provide convective heating and cooling of the home by allowing air to flow around the house. All around it. On the north side, the second wall is set 8 inches inside the outside wall, creating an airspace that is hidden from view. On the south side, a comfortable and practical sunspace exists between the heavily-glazed outer wall and the inner wall. Vents in the floor and ceiling encourage free airflow in a circular direction.

Here's how it works: In winter, when the sun is low, the sunspace—a 6-foot-wide room running along the entire south side of the main level where plants and vegetables can grow all year long—is bathed in direct sunlight. This warm air rises into large vents in the sunspace ceiling, flows under the roof and into the basement through the north-side airspace. As it flows, it imparts heat to the wood walls, floors, and ceilings—heat that is stored long after the sun goes down.

In summer, when the sun hangs high in the sky, vents located at the peak of the roof and at ground level on the north side allow for cool air to circulate around the house, pulling heat out of the structure and venting it out through the roof.

The house itself acts as the central heating and cooling system, and the thermostatic functions are achieved by way of vents, windows and channels that direct the flow of air in and around the house.

Other keys to Enertia homes' energy-saving design are the east-west axis (for greater passive solar potential) and the fact that each and every one is built into a hillside (for a walkout basement through which air circulates during sunny days and warm nights). The result is a solid wood house within an envelope of glass and wood that heats and cools itself passively.

Winter

Summer

Pretty much, anyway. Although it is rare for an Enertia home to require any form of active summer cooling, most of the homeowners I have talked to did occasionally resort to some form of winter heating after prolonged periods of cold, cloudy weather. A centrally placed woodstove is generally enough to do the trick, although one northern Alabama homeowner came up with a different solution: he uses an outdoor wood boiler to heat water for his basement's hydronic in-floor heating system. By using the home's own airflow circulation, he effortlessly heats the entire house.

Needless to say, Enertia homes are a good match for all types of renewable-energy systems. "To live in a house that heats and cools itself, and to live solar, you have to work at it," one homeowner told me. "It's a little different than walking over and turning up your thermostat. The beauty of it is, the house truly does work as it's designed to work."

— 3 —

Evaluating Your Electrical Demands

Watt(s) Your Use vs. Watt(s) You Need

To determine what size, or kind, of system to install, it will be helpful to calculate just how much electricity you currently use. This is quite simple—just divide the kilowatt hours (kWh) from your last electric bill by the number of days in the billing cycle. Unless you are quite conservative, or have a lot of watt-hungry tools and appliances, it should be in the range of 20 to 30 kilowatt hours per day.

That was easy. Now comes the hard part. Considering that the modules for a solar array will cost you somewhere in the neighborhood of $2–$3 per watt (and each of those watts will give you, on average, 0.003–0.004 kilowatt hours per day, which is to say that a 1,000-watt array will yield 3–4 kilowatt hours on a normal day at mid-latitudes), what are you willing to spend and/ or do without to get the most from your investment in renewables?

Below is a non-exhaustive list of things to avoid (like the plague, for those of you planning to live off the grid) if you want your new venture to produce satisfying results: air conditioning, electric clothes dryer, electric range/oven, electric water heater, electric heating (of any kind), older electric refrigera- tors/freezers (prior to 1999), incandescent light bulbs.

I'm sure for many of you the thought of getting by without air condition- ing is tantamount to doing without food and water. If you are building a new house, you can incorporate design features that promote passive cooling, as mentioned earlier in this chapter. But in many areas of the country even the best passive designs will offer little relief and living completely off-grid may turn out to be an untenable proposition. (In that case, I suggest you consider geothermal energy as the most energy-efficient cooling and heating technology available. See chapter 16 for more details.) Everything else on the list, however, can be substituted with an energy-friendly counterpart. You can just as easily use:

- Compact fluorescent light bulbs
- Gas clothes dryer
- Gas range/oven
- Gas and/or solar water heater
- Gas furnace/boiler
- Wood stove/boiler or masonry stove
- New, efficient, electric refrigerator and freezer

How much will these things lower your energy consumption? Plenty. If you trade out all the appliances on the first list (air conditioning, notwithstanding) with those on the second, you should easily cut

A Watts Up? meter is a simple way to measure the electrical usage of appliances (a toaster readout of 781 watts, or 0.781 kWh, is shown above).

your current energy bill by more than half (see the appendix for a list of appliances, and each one's energy usage, as well as a worksheet to calculate your energy usage). If you are serious about renewable energy, I suggest you invest in a Watts Up? or similar meter to measure the energy consumption of the appliances you now have, and then compare them to the appliances listed on the Energy Star website (*www.energystar.gov*).

Okay; so (on paper, at least) you've herded all the energy pigs out of your house and replaced them with efficient, state-of-the-art appliances. At last you've got your hypothetical energy usage down to a manageable sum. Maybe, you think, this renewable energy thing will really work. So what else can you do to maximize your investment?

The next best thing you can do is to give away all of your plug-into-the-wall electric clocks and replace them with battery-operated models. They're 100 times more efficient. A plug-in clock will use up to 26 kWh of electricity per year while battery-operated clocks will run for two years on a single AA battery. Go figure. After you change out your clocks, put everything that can draw a ghost load (a small load drawn by an appliance, even when the power is turned off) into a power strip that you can turn off whenever you're not using whatever is plugged into it. This includes TVs, VCRs, DVD players, cable boxes, and anything that has a little black AC to DC converter (such as laptop computers, chargers for cell phones and batteries, etc.). Even a load as small as one watt can add up over time: one watt drawn continuously for 24 hours is equal to running a 1,000-watt hair dryer or microwave oven for 1½ minutes each day.

You can be as militant or as charitable about this as you want, of course. We drew the line with the clocks on the gas range and microwave oven; hence, when I stumble down the stairs in the dark to throw another log in the wood stove I always know what time it is.

This will give you a good start. I've included other pointers throughout the book in the appropriate places. Once you invest in a renewable energy system and begin living within its framework (or constraints, depending on your point of view), you will forever amaze yourself with the little tricks you manage to come up with to save a watt or two.

Get Audited

A home energy audit is the best way to find out where the energy in your home is going and how to use energy more efficiently. Using cool diagnostic tools like blower door tests and thermographic scans, energy auditors will find your home's weak spots while determining the efficiency of your heating and cooling systems. They will also show you how to conserve hot water and electricity. Call your local utility and ask if they do professional energy audits. If not, they should be able to recommend companies that do.

MICK'S MUSINGS

Dogs work at less than 50 watts, and all of our energy sources are renewable. Think about that next time you take your TV for a walk.

---

one watt delivered for one hour = one watt-hour
1,000 watt-hours = one kWh (one kilowatt-hour)

Beyond Ohm's Law

The more you delve into the electrical particulars of renewable energy, the more you will hear the term "Ohm's Law" bandied about. This is curious, because electricians—not to be confused with electrical engineers, who use Ohm's Law on a regular basis—hardly ever have occasion to refer to George Simon Ohm's most notable accomplishment; at least not in its basic form.

Ohm's Law (*resistance* equals *volts* divided by *amperes*) is very handy for determining volts, amperes, and ohms (units of electrical resistance) when two of the three variables are known. If, for instance, you want to know the amperage of a circuit, you can measure the change in voltage across a conductor of known resistance. Likewise, you can use Ohm's Law to determine resistance and voltage.

But you won't. You are not concerned with the "R" part of the equation. That's the reason you have "wire size/line loss tables" in many renewable energy books (including this one). Thanks to Ohm's Law and the many formulas based on it, the phenomenon of resistance has been standardized in the electrical industry to the point that it is there without you even knowing it.

Let's put it in human terms: You want to brew a few cups of coffee in the morning. If the people who designed and built the coffee maker were ignorant of Ohm's Law, it would be a real hit-or-miss affair. But they knew all about it, so you don't have to. All you really want to know is how many minutes of direct sunlight it will take to fill your cup with stimulating black liquid. In other words, how are the volts (V) and the amps (I) required by the coffee maker related to the power produced by your solar and wind system? Resistance doesn't do you any good here—you're interested in watts (P). That is, after all, the units used to rate wind generators and solar modules. Fortunately for those of us who enjoy mental math, the formula is every bit as easy as Ohm's Law: **P = VI, or: watts (P) = volts (V) x amps (I).**

The above mentioned coffee maker runs on 120 volts, and draws 7 amps, so it will take 120 x 7 = 840 watts to run it. That's simple enough. So how much energy did it use? Well, if it takes 6 minutes to make coffee (and you don't leave the warming plate on after it's done brewing), then that's one tenth of an hour (0.10 hours, for those of you with calculators). So by taking 840 times 0.10, you see that it requires 84 watt-hours of electricity to scald the sleep out of your brain in the morning.

How much energy will the coffee maker use over the course of a year? If you divide 84 by 1,000, you will see that your coffee maker requires 0.084 kilowatt hours per day. Multiply that by 365 (days in a year) and you arrive at 30.66 kWh/year. This is a little more than two sunny summer days' output on a 2,400-watt PV system, meaning that over two days of each year are dedicated to making your coffee each morning.

There are two other expressions of this equation: $I = P \div V$ is the most helpful, since you can use it to determine wire size from the inverter or charge controller to the solar array or wind generator, among many other places.

On the other hand, since voltage is almost always a known variable, the final expression, $V = P \div I$, will probably be of little use to you or anyone else. You may find it ironic, then, that voltage will be the one immutable factor in the final configuration of your system.

The appendix gives formulas for calculating wire runs for various system voltages, plus a handy Wire Size/Line Loss table.

Power Formula

watts = volts x amps   OR   amps = watts ÷ volts

Example: 2 amps x 120 volts = 240 watts

WILLIE'S WARPED WITTICISMS

Dogs are energy pigs. A cat will give you 3 times the excitement at half the environmental expense. Besides, we're self-cleaning and fully operational in low-light conditions.

– 4 –

Grid-Tied or Off-Grid?
The First Fork in the Road

The desire to live off-the-grid is not universal in its urging. For some, it's an extension of a natural passion for self-reliance and simplicity; for others, it's simply the most cost-effective solution to a nagging problem. Most of us fall somewhere in the middle. We know people living within a stone's throw of a utility pole who nonetheless refuse to be hooked into the power grid. To each their own.

For LaVonne and I, it was undoubtedly cheaper to go with wind and solar than to tap into the neighbor's power line, over a mile away down a steep rocky road. Hence, we never bothered to ask the local power company what it would cost to run power to our new house. In any case, we figured we'd already given them enough money for one lifetime and suffered through enough blackouts to last through eternity. It was time for something new.

In 1999, we didn't realize the serendipitous nature of our decision. We just thought it was a cool idea. We didn't know that log homes (as our new home was) require far fewer natural resources to build than conventional homes, or that they are warmer in the winter and cooler in the summer. We only knew we wanted to build our own from the ground up and live in it.

Nor did we realize that, by powering our house with renewable sources of energy, we would, each year, prevent the consumption of over 7,000 pounds of coal on our behalf. This coal, when burned, would release nearly 15,000 pounds of carbon dioxide (CO_2) into the atmosphere; an accumulative tonnage of CO_2 that would, in less than 20 years' time, grow to outweigh a cube of solid concrete over 12.5 feet on a side.

We've both come a long way since then; this type of enterprise has a way of changing a person. We've learned a lot about energy efficiency and the utter practicability of homegrown electricity. We've learned new skills, and new ways to think about old problems. We've learned how to conserve when there is little, and how to better use whatever is in abundance. Best of all, we've learned that two people, working alone, can build a beautiful home, power it with the wind and sun, and live in it as though it were a palace.

Off-the-grid living is not for everyone, of course. In most areas of the country electric power is still relatively cheap and readily available, and many people live in homes with energy demands that preclude a wholesale conversion to renewable energy sources. It's hard to shell out hard cash for an expensive stand-alone renewable energy system when all the quiet and (usually) hassle-free power you'll ever need is right at your doorstep.

For those of you who are comfortably hooked into the grid, yet still find yourselves with this book in your hands, a partial conversion to alternative energy sources may be your ticket. But the question still looms: what is your rationale for using renewable energy? If you're worried that cheap, reliable power may soon be neither cheap nor reliable, your concerns are shared by the tens of thousands of Californians who installed some form of solar electric power after the rolling blackouts that have beset the state for the past several years. I can sympathize. The thought of being without electricity is a powerful motivator, especially in northern climes where electrical power, or the lack of it, can sometimes mean the difference between life or death. Having used my irascible Coleman generator to alternately heat two homes and pump water for forty horses during a five-day blackout following a spring blizzard, I readily admit to a deep-seated paranoia about not having electricity.

But even if your reasons are purely ideological, you'll still need to decide on which form of grid-tied system you want, though it shouldn't take too much thought. Basically, your choices are two: with or without. Backup batteries, that is. Both of these options will be examined in detail below.

PHOTO: NEVILLE WILLIAMS

Grid-Tie Systems: Marriages of Convenience

For those of you who do not plan to live way out in the boonies far beyond the nearest power line, but would still like to be on the front lines of the renewable energy revolution, a grid-tie system is the perfect way to get in on the action. Grid-tie systems are designed to run your house with homegrown power from a solar array and/or a wind generator, with power from the local utility company picking up the slack when your demands exceed your generating capacity. And, should your personal generating sources produce more power than your loads require, the extra power can be—in nearly all cases—sold back to the utility, who will then sell it to your neighbor, the one who does not share your foresight and highly evolved sensitivities.

Grid-tie systems can be configured over a very wide range of wattage and voltages. The system you choose will be largely dictated by your enthusiasm for the venture and the thickness of your pocketbook. No matter what size system you end up installing, there are (as I mentioned above) two basic types of intertie systems: those with batteries, and those without. If it's essential that certain electrical loads in your home continue operating when the grid power goes down, then you'll need to have some means to keep those systems operating. But are batteries the best way to go? That depends on where you live. If your house is in a place where power outages are infrequent and of short duration, then you might be better off with a direct-tie system and a propane-fired backup generator to keep your critical loads operational until the lights come back on. But if you live way out in the boonies where blizzards tear down power lines and you're usually the last on the list when the repair crews venture out into the field, then the cost of batteries may be justified.

MICK'S MUSINGS

Living off-the-grid is nice,
but I really miss the tasty treats
the meter reader handed out when
I'd bare my teeth and growl.

Grid-Tie Systems without Batteries

A grid-tie system without batteries is simplicity, itself. This is because the technology involved is so gratifyingly sophisticated that it requires virtually no effort or concern on your part to keep it working. (Did the word "virtually" raise a flag? It always does with me. Just the same, I used it because you still may want to sweep snow from your solar array, or adjust it for the seasonal movement of the sun. Hardly a big deal.) The heart of this system is the inverter. It will take power from your DC sources (solar array or wind generator) and turn it into an AC output in phase with that which is supplied by the local power company.

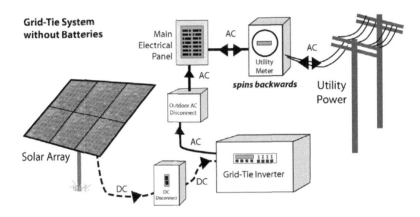

This type of system is far cheaper and easier to maintain than one that incorporates batteries, but it does come with one drawback: when the grid goes down, you go down with it. "But wait a minute!" you say. "What if the grid goes down in the middle of a sunny day? Don't I still get the power from my solar array?" No. Sorry. This is one of the many safety features built into grid-tie inverters. It's to keep hapless utility workers from getting unwittingly fried by the output of your vast solar array while they're working on the lines. Like it or not, it's a really good idea.

Grid Tied Systems with Batteries

Batteries in most grid-tie systems have it pretty easy compared to those in off-the-grid systems. And the less batteries have to work, the less work they are to maintain. So if the very word "battery" makes you cringe, don't let it. If you really want to make it easy on yourself (and who doesn't?), maintenance-free batteries are highly practical for battery-backup systems, since the batteries will have very little work to do as long as the grid is up and running. They can make your system pretty much trouble-free.

Besides the batteries themselves, an intertie system with a battery backup requires the addition of a charge controller between the DC source (PV array or wind generator) and the batteries. This is to condition and regulate the charge before it reaches the batteries. Extra safety components, such as breakers and disconnects, will also be needed, along with an AC subpanel.

These days there is really only one basic configuration in use for a battery-based system. Basically, the battery bank serves as a backup for when grid power is temporarily lost. You may be wondering how this works, since I just told you when the grid goes down your power goes with it, for the noble purpose of keeping power out of the lines that utility workers may be handling. Won't battery power go through the lines the same as power from the solar array? No, it won't. The reason it won't is as ingenious as it is simple: when the grid is down, your inverter will only provide power to critical circuits (such as the furnace, refrigerator, and a few lights) which will be isolated from the rest of the system for as long as the grid is down. These circuits will be located in a separate subpanel that the inverter keeps insulated from the outside utility lines. So, when the utility power is out, you won't have all the amenities you are used to, but at least you won't freeze in the dark while nibbling on rotting food.

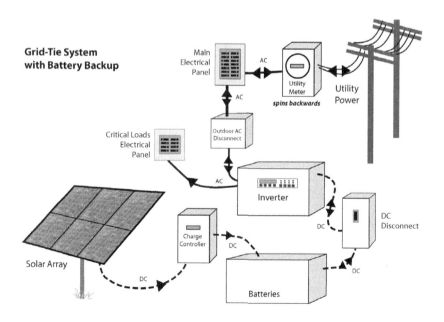

Grid-Tie System
with Battery Backup

Main Electrical Panel

AC

Utility Meter
spins backwards

AC

AC

Utility Power

Critical Loads Electrical Panel

Outdoor AC Disconnect

AC

Inverter

DC

DC

DC Disconnect

Charge Controller

DC

Solar Array

DC

Batteries

DC

During normal operation with the grid up and running, the inverter will first power the house loads with energy from your solar array and wind generator, and then charge the batteries with whatever is in excess. If the batteries are fully charged and the loads are taking less current than the charging sources provide, the excess will be sold back to the power company. Slick.

Net Metering

Which way should you go? Systems without batteries, while vulnerable to blackouts, are 3%–5% more efficient than battery-based systems, since there is no transition of electrical energy into chemical energy and back again. Generally, these systems are installed in places where the grid is fairly reliable, and rebates, tax credits, and other similar inducements are there for the taking. Battery-less systems are often viewed as long-term investments. By the time your system is paid off you will have enhanced the value of your home and reduced your electric bill to little or nothing, while creating an ironclad hedge against rising utility prices.

Another important factor is the price your local power company will pay you for your excess power. Most states have passed laws requiring most or all utility companies to buy back your power for the same price you pay for it. This is called net metering. If you live in an area where net metering is available, it makes sense to install a system that sells excess power to the utility during the day and buys it back at night.

Should you happen to live in a state where Time-Of-Use (TOU) billing is in effect, you will be able to sell your excess power to the utility for a premium during the day, then buy it back at a greatly reduced rate at night. It's like playing the stock market every day, knowing you'll be getting a generous return on your investment. The idea behind TOU was to reduce electric demand on weekday summer afternoons by raising rates between 1:00 p.m. and 7:00 p.m.

Now solar homeowners can sell kilowatt hours for about 30 cents in the afternoon, and buy them back for about 15 cents in the evening. It's legal, and the credit shows up as dollars on your bill. This has naturally become very popular with grid-tie system owners, because it will speed your payback by 15%–25% for most households. Time-of-Use favors the typical modern household where everyone's off to work or school most of the afternoon. You want to be sure to schedule electric chores before 1:00 p.m. or after 7:00 p.m., as much as practical. Pool pumps and air conditioning are good candidates for rescheduling. TOU billing has a summer schedule, May through October, when there's a large differential between peak and off-peak rates, which just happens to largely coincide with peak solar output in most locations. There's a much smaller rate offset the rest of the year, currently about 10 cents and 8 cents, but you still get to sell in the afternoon at a slightly higher rate than you buy back in the evening.

For more information about grid-tied systems, read *Got Sun? Go Solar*, a book I coauthored with renewable-energy pioneer, Doug Pratt.

Summary of Grid-Tie Systems

Grid-Tie without Batteries
- Requires fewer components than a battery-based system
- Is more efficient than a battery-based system
- Can sell excess power back to the utility company
- You'll be left in the dark when the grid goes down

Grid-Tie with Batteries
- Use sealed, maintenance-free batteries
- Can sell excess power back to the utility company
- Essential loads in your home still operate when the grid goes down
- Batteries need periodic replacement

Doug's Net Metering Setup

The 48 Uni-Solar peel-and-stick panels on Doug Pratt's garage are nearly invisible, but the 20 kWh per day of electricity (on average) that it provides is quite noticeable, as is evidenced by the zero electric bills from his utility company. In Doug's own words:

"We installed a battery-based system because lengthy power outages are not uncommon in our rural area, and with the nice California rebate we could afford to spend a bit more for the security. In my power room foreground you see the sealed batteries with a wooden cover so nothing can get dropped on them. On the wall, there's a pair of OutBack 3,600-watt inverters with a DC box and a pair of the superb MX60 charge controls on the near side, and an AC box on the far side for all the wiring and safety gear. Further down the wall there's a pair of standard circuit breaker boxes. One is for utility power, the other for inverter power. Note they're connected by a gutter box so individual circuits could be easily moved if we change our mind about what wants backup power.

"The passive-solar, geothermal cooled and heated, all-electric retirement home we built to go with this admittedly oversized shop was burning a little more power than we were producing. The electric bill was $110 per year. Can't have that! A new 5.0-kilowatt ground-mounted, direct grid-tied PV array that's aimed southwest to take advantage of TOU billing, has covered our shortfall, and leaves enough extra to charge an electric car for the 5-mile trip to town."

— ∫ —

Where Does Everything Go?

Finding Space for Your Personal Power Company

Every house has a mechanical room, where the furnace or boiler, and the water heater are located. If you are building a new home back in the woods and plan to pump water from a well, you may also need to allow space for a cistern or a pressure tank. Basements are great places to hide all the stuff you don't want to look at. As you plan your mechanical room, consider what systems you might want to use: a solar hot-water storage tank, perhaps, or even a geothermal heat pump. Allotting space at the outset is ten times easier than trying to make space later after it's all filled up with treasures and trifles.

Electrical Room

In the house where you are now living, there is a single conduit with three heavy wires running into your main electrical panel. This conduit runs from the utility grid to the outside of the house, through a wall, to the panel. It's hidden in the wall and takes up no living space. In grid-tied systems without batteries it's pretty much the same; the inverter and other components are usually mounted to an outside wall and thus take up no interior space.

Things are a little different in a grid-tied or off-grid home using batteries. If you are planning a renewable energy system *with* batteries, your house will also need to have an electrical room, or at least some out-of-the-way place big enough to contain all the components of the biggest solar/wind system you will ever conceivably build. This may include wall space for the inverter(s), DC disconnect(s), charge controller(s), perhaps a 120-240 volt AC transformer, and floor space for your battery bank.

To give you a rough idea of how much space everything takes, our electrical room is 9-feet by 4-feet. Our two banks of 12 L-16 style batteries each fit nicely in a space 48 inches by 40 inches. By squeezing components closer together we probably could have gotten by with a little less space, but not much. If you can, leave yourself plenty of extra room. In particular, allow yourself space to expand the size of your battery bank, should you want to add more batteries later on.

Most wind and solar suppliers/consultants will be more than happy to help you design your system, and their catalogs give precise dimensions of every component you'll need in your electrical room, including the batteries. So, once you familiarize yourself with the essentials of a photovoltaic solar and wind system and understand what's what, it's a simple matter to draw a basic diagram detailing where everything will go.

A few other things you'll need to keep in mind:

- **Your battery box cannot be located directly beneath any serviceable component**, such as an inverter (for servicing and safety reasons, according to the National Electric Code). And neither can anything else. It's really too bad, because that's the logical place to put the batteries, since you will want the large cables connecting the batteries to the inverter to be as short as possible. One clever way around this problem is to build the battery box on the *other* side of a frame wall, just *behind* the inverter, and running the heavy cables through the wall. Conversely, locating the battery box just to the side of the inverter will make for a short cable run.

- **Flooded lead-acid batteries—the ones most used in off-grid systems—need to be in a sealed box, vented to the outside.** It only takes a one-inch PVC vent pipe, but the closer you can locate the batteries to an outside wall, the better.

- **Inverters may hum.** *How* much they hum depends on the brand. To keep your sanity, put the inverter in a room that can be sealed off from the rest of the house in general, and your bedroom, in particular.

- For the sake of efficiency, it's best to **locate all of your solar and wind**

Where is North?

Those of us deficient in surveying experience generally rely on a compass and a chart giving the magnetic declination for our particular area to find true north. If you're really lucky, you might be able to get within a few degrees by this method. If that is acceptable, great. After all, experience has shown that plus or minus five degrees in the alignment of the array makes virtually no difference in the output of power.

But why settle for an approximation with a fractious compass when you can get a nearly perfect alignment with a couple of T-posts and the ability to locate two of the most conspicuous constellations (the Little Dipper and the Big Dipper) in the night sky? Polaris, the bright, terminal jewel in the handle of the Little Dipper (officially known as Ursa Minor) is the North Star. It is the pivot point around which every other star in the night sky revolves. The two bright stars delineating the outer bowl of the Big Dipper (Ursa Major) point to it from 30 degrees away.

Once you know where Polaris is, all you need to do is drive a post into the ground near where the array will be situated. Then, as you sight directly over the top of the first post, have someone to the north of you move a second post east and west until a sight line along both posts points directly to Polaris. You now have a true north-south axis to work from for a perfect alignment of your solar array.

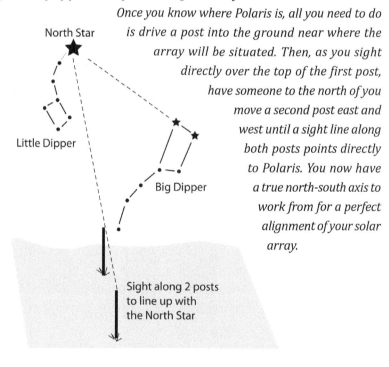

North Star

Little Dipper

Big Dipper

Sight along 2 posts
to line up with
the North Star

components in the same room as the main electrical panel. That way everything is in one handy place.

- **All serviceable components (such as inverters) need to have at least 36 inches of space in front of them.** This is to allow plenty of room for defensive tackles to work on them.

Southern Exposure for the Solar Array

Our house is situated on a saddle 500 feet above a creek and is surrounded on all sides by distant mountains. The sun rises here about an hour later than it does on the plains to the east, and sets an hour earlier. It's about as good a spot as you can hope for in this area. Since the solar wattage available from the early and late-day sun is only a small fraction of what's available at midday, we lose practically nothing to the surrounding hills. At 40 degrees north latitude we have all the sunshine we need to power our house, even in late December when the days are so short that lunch is the only meal we can enjoy in daylight.

Since your solar array will be a major investment, you'll want as much direct sunlight as you can get. If you plan on building in a valley, be absolutely sure your winter sun isn't blocked by the hillside to the south. This can be deceptive. In July, when the sun burns high in the sky like a celestial blast furnace, it's hard to recall just how far south old sol retreats in winter. Several years ago I rented a little place in a deep valley one canyon north of where LaVonne and I now live. It was bright and sunny in August when I moved in, but as fall approached I watched with dread as the sun crept ever south. Then, early in November, it just disappeared and didn't offer up a single yellow ray until the following February. The sun's arc was nearly identical to the topography of the mountain to the south; one day the sun was there and the next day it wasn't. Fortunately, I didn't own the place and it was wired into the grid, so I

Do you want to know for certain how surrounding trees and hills will affect your array? The Solar Pathfinder shows where the shadows will be any time of the year.

Our ground-mounted solar array is easy to clean and can be adjusted to meet the angle of the sun. The steep winter setting is shown here, as is a nice deep snow.

could at least see what I was doing as I shoved copious quantities of cordwood into the stove to keep from freezing to death during the three months of wintry shadow.

Plan ahead where you're going to put your solar array. The roof of the house is fine; the modules are out of way, they're safe from the depredations of deer, elk and your neighbors' cows, and your view to the south will be unobstructed by your array's unflattering backside. But modules are much easier to clean, adjust for seasonal changes in sun angle, and sweep the snow from if they are mounted on the ground or on the side of a deck. And when it comes time to re-roof your house you won't have to disassemble the entire array. If you do choose to go with a ground-mounted array, however, you won't want it much more than 100 feet from your battery bank, because in low-voltage DC systems—as is common with battery-based setups—the size of wire required to carry the current without substantial line loss increases greatly with distance, and heavy wire can get pricey. It's not much fun to work with, either.

Thinking About Wind

I'll have a lot more to say about wind in a later chapter, but it might be helpful to touch on a couple of points here, just to get you acquainted with the subject.

The wind, of course, blows where it will, and you will either have enough wind to justify the expense of a wind turbine and a tower or you won't. With solar prices plummeting and turbines continuing to get more expensive, you really should live in a very windy place to even consider putting up a wind turbine. Some states are windier than others, and every point within each state is different from every other. Moreover, some months are windier than others, as are some years. Secluded valleys are generally not good places for wind generators; mountaintops are great. You could get a ton of expensive equipment and monitor the wind at your site over the course of a year to determine if it blows enough to justify the purchase of a wind generator, or you could follow my simple rule of thumb: *If the wind blows hard enough and often enough to annoy you, you can probably make good use of a wind generator.*

Once you determine—by whatever means—that you do have enough wind to make a wind-generating setup viable, you'll need to start thinking about where to put your tower. This is a little trickier than finding a spot for a solar array, since it will need to be a safe distance from the house (if it's a big turbine, which is to say, over 500 watts), but not so far away that you need to take out a loan to buy the wire leading to it.

Furthermore, it will have to be high enough off the ground that nearby buildings and trees do not block or dampen the wind from any direction. Ideally, the generator should be mounted at least 30 feet higher than the highest point within a lateral radius of 300 feet. This isn't always possible, but it's a starting point.

Monitoring Your System

How do you know how well, or poorly, your shiny new PV system is working? That would be the system monitor, and there are several choices over how this data is delivered to you.

Inverter Display Every inverter comes with an onboard display that will show how many watts are being output at the moment, how many watt hours have been delivered so

Mate3, a remote display by Outback Power

far today, how many kilowatt hours have been output since first being turned on, and if there are any problems, it will display the fault code(s). Plus, there are probably a few more readings the engineers threw in the rotating display. For many folks, this is sufficient. The display comes with the inverter, and all it costs you is the effort to walk out to the inverter and look. But sometimes that's difficult or impossible, especially for grid-tie inverters mounted outside the home. Which is why we have more choices.

Remote (Hardwired) Display Most inverter manufacturers offer a communications port that will send data to a PC computer within 300 feet. This comm port is standard equipment with some manufacturers, others charge extra. The software to read, display, and manage the data is universally a free download from all manufacturers. The CAT5E (Ethernet) cable to get from your inverter(s) to your PC can be simple or complex depending on the distance and tidiness required. The primary disadvantage of this monitoring type is that the PC must be turned on to receive the data. The inverter doesn't save up data till the computer is ready, it simply transmits data as it happens.

Remote (Wireless) Display Some inverters offer a wireless version of a remote display. These are separate stand-alone displays that don't require a home computer. Cost is in the $400–$500 range. A small transmitter and antenna is located on the inverter; the receiver must be within about 150 feet. If there's concrete, rebar, brick, or rock in the way, that distance can be much less.

Web-Based Monitoring Do you want to see how well the PV system on your vacation house is doing? Or if the battery-based repeater on some remote mountaintop is still working after the last storm? Web-based monitoring makes it easy. Every major inverter maker offers some way to plug their unit into your internet router, and several offer free or very low-cost sites to host, display, and store that data for you. Premium monitoring sites feature some automated analysis tools that will detect problems and send email or text warning messages. These are being widely used by better solar dealers as a service/maintenance perk. Your dealer will know if your system is under-performing before you will, and can possibly provide repair services before you're even aware there's a problem. Prices vary a great deal for web-based monitoring packages.

WILLIE'S WARPED WITTICISMS
Cats spend hours a day monitoring the Dog's output (which, frankly, isn't much).

— *6* —

Sizing and Pricing Your Solar Electric System
Where Dreams Meet Reality

Choosing a System Voltage for Off-Grid Systems

Not too long ago, choosing a system voltage—the voltage at which the inverter and battery bank operate—began with the solar array. All modules were either 12 or 24 nominal volts so almost all PV systems were wired for 12-, 24-, or 48-volt operation. And while 12- and 24-volt PV modules are still available, they are becoming rarer by the day. This is because the solar industry has become almost completely standardized around the 60-cell PV module (think 20 nominal volts; 30 volts maximum). The typical module will be 220–260 watts depending on the make and model. Fortunately, we now have smart MPPT-type charge controllers to down-convert the power from series strings of three 60-cell modules into power the batteries can happily digest, so the array voltage is no longer tied to the system voltage.

The system voltage, then, is determined by the inverter. The batteries you buy will be 2, 4, 6 or 12 volts each and they can easily be wired in series to match up with the inverter. So what inverter voltage should you choose?

If you are looking for a catch, it isn't in the price. There is no difference in cost between a 12-volt inverter and a 48-volt inverter. If you want to include a few 12-volt DC circuits in a 24- or 48-volt house, you'll have to buy a step-down converter to pull the lower voltage out of a higher-voltage system, but the money you save by having a higher-efficiency system will pay for that minor item several times over.

The only real downside in systems over 12 volts is in the multiples of batteries you will have to buy if you decide to add more later. With 6-volt batteries, 12-volt systems can be enlarged in twos; 24-volt systems by fours, and 48-volt systems by eights. It can get expensive very quickly. On the other hand, all electrical systems operate more efficiently at higher voltages. We opted for a 24-volt system years ago, when we knew a lot less about this business than we do now. It was a choice that has worked well for us and I don't imagine we'll be upgrading to 48 volts anytime soon. That being said, most battery-based systems these days are designed to operate at 48 volts and if you are satisfied that your battery bank is properly sized from the get-go, I would heartily recommend going with a 48-volt system.

How Much Generating Capacity?

Our home runs with power from the sun and wind; most of our neighbors are strictly solar. Maybe you'll be lucky enough to add microhydro power to your equation. After you read the next few chapters about solar/wind/hydro equipment, you'll have a better idea of what will work for you. You'll also want to study the detailed appendix with worksheets, references, and resources.

....

How Much Power Will My System Really Deliver?

A fudge factor of 70% takes into account all the real-world effects of dirty modules, dirty air, high humidity, hot modules, wiring losses, small bits of shading, inverter inefficiency, and all the other little things that are less than laboratory perfect in a working system. For example: 3,060 watts x 5.5 hours x 0.70 = 11,781 watt-hours or 11.78 kWh per day. Remember, this is a yearly average. Most folks will see about twice as much power production in the summer as in the winter. For battery-based systems, a factor of 65% is closer to reality. If your array is perfectly shade-free, installed properly, and you're in a dry climate above 6,000 feet elevation, then bump the factor up an extra 3%–4%. So long as there are no serious shading or other performance-reducing site problems, this formula usually yields a conservative estimate. Most customers report slightly better performance with their actual installed systems. The long-term Redbook charts state, "Uncertainty ± 9%." Take your result as a very educated guesstimate—your mileage may vary—but probably not more than 5%.

....

How Many Batteries?

While one of the most common problems with off-the-grid renewable energy systems is too few batteries, it is also possible to have too many. Why? Because it's not always enough just to keep the batteries in a so-so state of charge. The batteries really should be brought to a full charge regularly, especially if you are using a meter to monitor the batteries' state of charge, (meters can begin to lose their calibration after a few days unless the batteries are charged often). More than that, however, lead-acid batteries need to be equalized—over-charged under controlled conditions—every so often to ensure that the plates remain free of sulfates. Of course, many people use a generator for equalizing their batteries, but you shouldn't have to with a properly sized system. Besides, it's cheating.

The point is, your battery bank needs to be sized in accordance with your loads and your generating capabilities. With too few batteries you will be wasting sunlight and wind; with too many, the batteries themselves will suffer.

If you size your off-grid system optimally the first time out of the gate, your success may well be attributable as much to serendipity as to math. Every off-grid family we know has added onto to their system over the years. The key to sizing your system, then, is to leave yourself room to add on later. This means starting with the right inverter. It means running heavy enough wire to the array that you can later add more modules without having to worry about line loss. And it means leaving yourself space in the electrical room—and the battery box—to add more batteries, if need be.

See also chapter 11 and the Battery-Sizing Worksheet in the appendix.

How Many Watt-Hours in a Battery?

To convert a battery's amp-hour capacity to watt-hours:

amp-hours x voltage = watt-hours

Example: 390 amp-hours (L-16 battery) x 6 volts = 2,340 watt-hours

Discharge to 50%: 2,340 x .50 = 1,170 watt-hours available

Sizing Grid-Tied Systems

Direct Grid Tie (no batteries)

A direct grid-tied system without batteries is really nothing more than your own little piece of an unimaginably large power grid. And as the grid goes, so goes the direct-tied system wired into it. Your system will not provide you with any power unless the grid is up and running, so how big or small you make it is really just a matter of preference. Why are you installing this type of system? If it's because you want to pitch in and do your part to offset the burning of tons of coal, or because you just aren't comfortable knowing that countless atoms of Uranium-235 are being disintegrated on your behalf, then your decision is a simple monetary one: buy whatever you can afford. On the other hand, if you hope to see a credit on your bill from the power company now and then, you'll need to do some figuring.

Start by dividing the monthly kWh total on you power bill by the number of days in the billing cycle to arrive at your daily consumption. The average American home is in the range of 20–30 kWh per day; let's cut you some slack and say it's 20. If you wanted to average 20 kWh/day of solar energy production here in sunny Colorado you'd need an array with a nominal rating of around 4,500 watts—approximately double the size of my array. If you live in the Northeast, you might need twice that much to completely offset your current power consumption.

This raises the question of where to put it all. A tightly packed 9,000-watt array will take up over 700 square feet of space, either on your roof or in your yard. If this sounds excessive, then perhaps you can find ways to cut your power consumption, as suggested earlier. Or you can opt for a somewhat smaller system that reaps a credit in summer that will be used up (and then some) in winter. In any case, you will have greatly reduced your power bill, especially if you live in an area where daytime rates are more expensive that nighttime rates, since you will be selling high and buying low.

Grid-Tied Systems with Backup Batteries

Are you contemplating a grid-tie system with batteries? If so, the planning stage becomes more critical, since you will need to know : 1) How much power is consumed by the appliances and home systems you wish to keep operating when the grid goes down (these are your critical loads and will be broken out into a separate subpanel); 2) How many watts of solar capacity it will take in your particular geographical area to keep the batteries at an acceptable state of charge during a power outage; 3) How big of an inverter you will need to provide power to the loads it will be running; and 4) How many batteries will be required to power your home during a power outage.

You'll also need to figure out where you're going to put all the components, since batteries take up a fair bit of room and need to be as close to the inverter as possible. On the plus side, the sealed batteries you'll be using can be stacked on a sturdy rack and stored in any position, meaning that they will take up a lot less space than off-grid, wet-cell batteries that have to be readily accessible for maintenance.

The difference between sizing a battery-backup system rather than an off-grid system is that with the latter you are dealing with a known quantity. Off-gridders need systems that will power everything, all the time. But a battery-backup system is a little harder to nail down. Do you want it power your critical loads during a weeklong power-pole-snapping blizzard while your solar array is buried under feet of snow, or are you merely needing battery backup for an hour or two every time the grid has a little hiccup? Most power outages are of relatively short duration; really big ones are a lot more rare. Does it really make sense to spend thousands of dollars on extra batteries that may well live out their 8- to 10-year lifespan without ever once being called upon to assist you through a blackout your grandkids will tell their grandkids about?

A smarter idea is to plan for an average blackout and spend the money you would have spent on those extra Armageddon batteries on a gas generator, instead. It will probably cost about the same and outlive the batteries by a considerable margin. With a generator *(see chapter 14)*, you can charge up the battery bank when your solar array is otherwise encumbered, and if you go about it right you can make the whole system automatic and trouble free.

Whatever type of system you may be thinking about, the appendix has a number of worksheets to assist you through the planning stage. Even if you intend to have a professional installer do the work, the worksheets will give you a glimpse of what you're in for.

Perhaps the most useful solar module I own is the one that runs my little workshop. Completely independent from the house system, my detached workshop is powered by a single 165-watt module, eight T-105 batteries, a Xantrex C-40 charge controller and a no-frills Aims 2,500-watt modified-sine-wave inverter. I use this small system for grinding, cutting, sanding, drilling and even welding, on large projects and small. The key is that I don't use it every day, and generally not for more than an hour or two on the days I do use it. It gives the batteries plenty of time to refresh themselves between duty cycles and ensures that I will never draw more from them than they have to give.

···

How Much Will It Cost?

Off-Grid and Grid-Tied Battery-Based Photovoltaic Systems

What does it cost to piece together a new solar-electric system? Over the years that's been something of a moving target. LaVonne and I paid $4.80 per watt for our first six Solarex modules back in 1999. At the time there were only a handful of manufacturers. Today there are probably more companies making solar modules than tennis shoes, which means there is currently a glut on the market. High-quality PV modules can easily be found for less than $2.00 per watt, which makes it a lot easier to start off with a fairly large system.

Batteries, on the other hand, have nearly doubled in cost in the last 10 years and every indication is that prices will continue to rise. Inverters, charge controllers and other electrical and electronic components have held fairly steady with the rate of inflation, with the caveat that they have, by and large, increased in sophistication over the years and such refinement always comes with a price tag.

To buy a complete off-grid system comparable to ours (around 2,300 watts), you should count on spending $8,000–$9,000 for the modules and electronics, and another $3,000–$6,000 for the batteries, depending on what brand and how many you buy. Throw in another $1,000–$2,000 for wire, mounting hardware and other sundries. So roughly speaking, you can buy all the parts, materials and components you need for a very respectable off-grid system for anywhere from $10,000–$12,000. You can creep closer to the low end of that estimate by shopping around. Dealers often overstock solar modules and let them go at fire-sale prices if you are willing to buy them by the pallet. You can also help yourself out by doing little things like building your own module mounting system, if you're up to the challenge.

What applies to off-grid systems also applies to battery-backup systems. The main difference here is that you will need fewer batteries at more expense per battery. But unless you want a really robust backup system, you should be able to save some money here over an off-grid system.

Direct-Tied PV

Direct-tied PV systems are considerably simpler than battery-based systems, but as a rule they are much larger than their off-grid counterparts. This is largely because off-gridders are invariably more parsimonious about their energy usage than their grid-tied friends. For a turn-key, direct-tied system where your only involvement is to sign the contract, figure on spending around $1,000–$1,500 for every kilowatt hour you hope to knock off your electric bill each day. An average 3.0 kW system will run about $15,000–$20,000, installed, depending on the difficulty of the installation.

Installation Costs

Installation costs are generally on the rise. Most solar installers have worked hard to earn their certifications and they rightly expect to be compensated for all the effort they put into it. In northern Colorado labor costs are upward of $100/hour in many areas but, like most other commodities, labor is only worth what the local market will bear; it may turn out that you can get a quality installation for less money in your neck of the woods. In any event, plan on paying at least of $2,000 in labor to have your system installed.

Should You Install Your Own System?

In case you haven't guessed, I'm a hands-on, do-it-yourself guy. I can't bear to pay someone to do something I can do myself. It's an attitude that often gets me into trouble but, hey—that's me. So it may seem disingenuous of me to suggest that it might be in your better interests to hire someone to install your system. After all, I'm sure you know your own limitations better than I do. But if you are planning a system that hooks into the power grid, I seriously suggest you consider having it done by a professional installer. Why? Two reasons. Number one, you'll be dealing with lethal doses of electricity; if you make one mistake you may not be around to make a second one. And number two, many rebate and tax credit programs carry a self-install penalty.

In other words, if you do the work yourself you'll feel it in the wallet. It's a great incentive to kick back and watch someone else do the work.

PHOTO: NEVILLE WILLIAMS

Solar Leasing

If you would like to run your home all or in part with direct-grid-tied solar electricity but do not wish to pay the upfront costs or bear the responsibility that comes with owning a solar-electric system, you can lease the equipment at a price guaranteed to be lower than your current electric bill, with zero costs to you. All you have to do is sign the contract.

Solar leasing is an idea that is catching on in many areas, although currently it is only available in a handful states. The basic idea is this: once you sign the lease agreement—usually for a 15- or 20-year term—the leasing company goes to work. They apply for all the necessary permits and, of course, rebates and other incentives. They then install their system on your roof, where it quickly begins to dump valuable solar electricity into the local power grid. For their trouble, they get all the rebates and tax credits that would have gone to you, plus a monthly lease payment from you (as laid down in the lease agreement).

The leasing company, in turn, guarantees that their system will produce a given amount of electricity during a given month, whether or not it actually does so. You pay only for the electricity that you use over and above what the leasing company claims their system will produce. Plus, of course, the cost of the lease payment. What this means for most homeowners is a modest savings over what they would have paid for electrical power had no system been installed.

Is leasing a good deal? It depends on a number of factors, but mostly on your current financial situation. Certainly if you can afford to buy a system outright it will pay for itself long before the term of the lease expires, after which time your system is like money in the bank for decades to come. At the end of a lease agreement, by contrast, you still have no equity in a 15- or 20-year-old solar system, and if you wish to keep it on your roof you will then have to buy it from the leasing company for a considerable fraction of its original cost.

There is also the problem you will face if you decide to sell your house. If you own the system outright it will certainly enhance the value of your home. But if you are leasing a system it may turn into a liability if the buyer is not comfortable with the idea of assuming your lease. Should this be the case, you will either have to buy out the lease or, if feasible, have the system moved to your new home—at your own expense.

_– 7 –_

Solar (Photovoltaic) Modules

Sunshine at Your Service

The first time I ever saw a polycrystalline solar module from two feet away I had the feeling I was looking at something magical. Deep inside, it looked like thin slices of the finest Persian lapis, overlaid with semi-transparent crystalline mirrors tinted in a hundred hues between blue and purple. It was so awe-inspiring it might have been a master jeweler's creation, except that closer to the surface lay a thin grid of aluminum channels; pathways for the energy with which this strange creation resonated. Were such a jewel of modern alchemy to travel back through time—to ancient Sumer or Babylon, perhaps—it would've changed the course of civilization. Certainly it's beginning to change _this_ civilization.

No one ever intended solar modules to be so beautiful, of course; they just turned out that way as a direct result of their design. Made from the basest of materials—nothing more special than thin layers of silicon (the stuff beaches are made of), coated (doped) on either side with atoms of dissimilar electrical properties—they take on their other-worldly appearance and their special properties through a laborious and painstakingly precise manufacturing

MICK'S MUSINGS

Pound for pound, dogs are twice
as energy-efficient as cats.
If this sounds high, calculate a
feed bill for 50 pounds of cats.

process. The end result is a multi-layered substrate through which electrons are set in motion after being jostled by sunlight.

The phenomenon of converting sunlight into electrical energy was first observed more than two decades before the outbreak of the Civil War, but it wasn't until the 1960s and the Cold War that the first efficient solar cells were produced for use in the space industry. Though exceedingly expensive, by 1970 more solar modules were produced for use on Terra Firma than in space. Since then, the use of solar modules has steadily increased around the world, while the price has been driven down from Buck Rogers' price of over $1,000 per watt to less than $2 per watt for contemporary earthlings. Still a little pricey but no longer out of reach.

Types of Solar Modules

The primary energy producing unit of any photovoltaic system is the cell. Crystalline or polycrystalline, silicon cells each produce about 0.50 volts, regardless of how large or small they may be. Bigger cells produce more amperage, but the voltage remains the same.

To increase the voltage to the point where it can be used to charge a battery, several cells are connected in series and arranged together within a module. Also commonly known as a "solar panel," a module is one of the glass-covered, aluminum-framed units you buy from the dealer and install on or near your house.

Most 12-volt residential modules on the market today actually range from 16.5 to 18 volts. This may seem high for a 12-volt system but, like water running downhill, it's necessary to keep the current flowing from the module to

Not all PV modules are square. PHOTO: SHARP

the battery rather than the other way around. Solar cells experience a drop in voltage with increased temperature (during the summer months, for instance), and batteries in a 12-volt system often reach 15.5 volts or more while in the process of equalizing. If the module voltage were any less, there wouldn't be enough electrical "pressure" to bring the batteries to a full state of charge.

Modules are rated under optimal laboratory conditions—direct midday sunlight, at a cell temperature of 25°C (77°F). Performance drops off with a change of sun angle, a rise in temperature (each one-degree-Celsius rise in temperature will cause a 0.5% drop in voltage) or, of course, the appearance of clouds. Common sizes for modules in residential arrays range from around 100 to 240 watts, though somewhat larger modules are available. While it would seem that you would be able to get more watts per dollar by buying bigger modules, this is not always the case. Prices fluctuate from dealer to dealer, week to week. If you're not too picky about size, you can usually find some good deals. If you do buy smaller modules, however, you need to real-ize that they will entail more wiring per kilowatt, as well as more work and material for the mounting frame.

Most solar modules in use today are made from crystalline or polycrystal-line silicon, protected by a layer of tempered glass. They work well for most people in most applications, but they are not without certain shortcomings. As I mentioned above, the voltage begins to dwindle as the cells heat up. In addition, performance drops off considerably under conditions of partial shading, as often happens when snow slides off one half of a module but not

---

Actual Module Output vs. Rated Output

The watt rating (Pmax) of a solar module is the product of its rated amper-age (Ipmax) and its rated operating voltage (Vpmax). A 120-watt module might produce 7.10 amps of current at 16.9 volts (7.10 x 16.9 = 119.99). But when the batteries draw the voltage down to 12 or 13 volts there is not a corresponding rise in amperage, and a significant amount of power is lost (7.10 x 13 = a mere 92.3 watts). How do you get it back? Power point tracking can help. For more on this watt-saving technology, read the section on MPPT Technology, chapter 8 and the appendix.

the other. In this case, the clear side tries to run current through the shaded side, to no avail. Since the module voltage is the sum of all the individual cell voltages, a partially shaded module may not have sufficient voltage to charge a battery. The same thing occurs when one of two modules wired in series is shaded. In systems over 24 volts, damage can occur to the shaded module when the unshaded module tries to drive it. For this reason, module manufacturers generally install bypass diodes to avoid damage to modules operating at voltages higher than 12 volts.

Finding a Place for the Array

While it would seem logical that the roof would be the best place to locate your array, there are some drawbacks to rooftop mounts that you should consider. First, it's difficult to sweep off the snow. In the winter, when the mercury hovers around zero the day after a snowstorm, the snow may cling to the array all day, even if it's bright and sunny. You could be charging at full power, but you won't get any power at all if your modules are buried under 6 inches of snow. Even on foggy days following a storm there's lots of power to be had—as long as you can remove the snow.

Additionally, rooftop-mounted modules are hard to clean in the summer after a little rain sprinkle leaves them looking dirty. And, of course, with a roof-mounted array it's practically impossible to adjust the array for seasonal variations in the sun's angle.

That being said, not everyone has a big sunny yard or a south-facing deck, so for you the roof may be the only option. Fortunately, roof-mounted

Popular with off-grid homes, pole-mounted arrays can be easily tilted for seasonal adjustments.

arrays no longer need to be the tedious and potentially leaky affairs they were in years past, since commercially available standoffs can now be painlessly incorporated into your home's roofing system. While the standoffs themselves do in fact penetrate the roof deck, they are made waterproof with standard plumbing vent jacks. Thus they will hold your array safely off the roof while virtually negating the possibility of leakage. For standing-seam metal roofs, non-penetrating mounts can be used. Designed to work with stout, easily adjustable aluminum frames, standoffs take the guesswork out of roof-mounted arrays.

If mounted on the ground, the array ought to be as close to the house as possible in a location where it can receive full, unimpeded sunlight for the three hours on either side of noon. Bushes or trees that cast a shadow against any part of the array can greatly reduce the energy output, even in winter when the branches are bare. You should either remove the foliage or find a more suitable spot to place the array, even if that means the roof.

Arrays should always point due south, as long as there are no peripheral obstructions. Trees or hills to one side of the array may make it more advantageous to position the array at an angle that makes the most use of what sunlight there is.

If the best spot to locate the array is more than 50 feet from the house, you might consider designing your system at a higher voltage, since this will greatly decrease the diameter of the wire needed to carry current to the house with a minimum loss of current through the line.

Tilt Angle of the Array

The sun passes through 47 degrees of arc twice a year as it treks from the Tropic of Capricorn to the Tropic of Cancer and back again. At 40 degrees north latitude (as we are here in northern Colorado) the sun moves from 26.5 degrees above the horizon at noon on the winter solstice (December 21), to 73.5 degrees at noon on the summer solstice (June 21).

Obviously, a solar array that is perfectly perpendicular to the sun will produce the most power. If you diligently adjusted your array every few days, you would want to set it at 90 degrees minus the sun's angle. The array on the first day of winter would be set at 63.5 degrees (90 minus 26.5 degrees), and gradually adjusted to the first day of summer at 16.5 degrees (90 minus

73.5 degrees). But you won't adjust the array angle every few days, of course, unless it is tantalizingly easy to do so. (Keep reading...)

Shortly before we handed our electrician his walking papers, he brought a solar installer to our house in hopes of getting some advice on how to wire our system to code. The installer walked down to our cabin to check out my handiwork with that basic system. When he saw the array there set at 40 degrees, he told me the angle was way too low—we should set it at 55 degrees and leave it there. I thought the man was daft. Then I realized that his logic was shared by almost everyone around here. This is how it goes: if you set your array to capture the most light during the coldest, darkest time of the year, then whatever percentage of energy you lose in warmer months will be more than compensated for by the extra hours of sunlight.

This line of reasoning is based on the fact that there are about twice as many hours of daylight in summer. It is also assumed that cloud patterns are essentially the same throughout the year. But that's not always true. In Colorado, for instance, the sky is generally clearer in winter. In summer, though it doesn't often rain, the clouds roll in over the western mountains practically every afternoon, greatly diminishing the amount of light that falls on the solar array. Couple this with the fact that the sun rises and sets so far north in the summer that the modules need a low angle to capture any early and late day sun, and the argument for a steep, stationary angle falls apart.

That being said, I have to admit that most of the dozen or so people we know who live entirely off the grid have their arrays set at a steep angle and never bother to adjust them for the seasons. It's probably the reason why LaVonne and I are often treated to a chorus of generators on warm summer nights, while our generator is resting comfortably in the garage. The bottom line? If you plan on a ground-mounted array, make or buy an adjustable frame. Then you at least have the option, whether you use it or not.

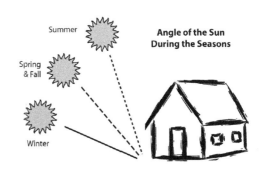

Angle of the Sun During the Seasons

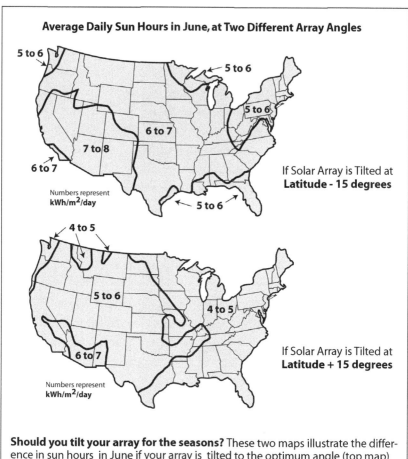

Average Daily Sun Hours in June, at Two Different Array Angles

5 to 6

5 to 6

5 to 6

6 to 7

7 to 8

6 to 7

Numbers represent
kWh/m²/day

5 to 6

If Solar Array is Tilted at
Latitude - 15 degrees

4 to 5

5 to 6

4 to 5

6 to 7

If Solar Array is Tilted at
Latitude + 15 degrees

Numbers represent
kWh/m²/day

Should you tilt your array for the seasons? These two maps illustrate the difference in sun hours in June if your array is tilted to the optimum angle (top map)
versus if it is kept at the winter setting (bottom map).
Data from the NREL Resource Assessment Program.

Reckoning Where the Sun's Going To Be

If you choose to adjust the angle of your array, it's a very simple matter to measure the sun's angle. All you need is something that casts a shadow, like a deck
railing, a fence post, or even the backside of your solar array; a long, straight
board; and a floating-pointer angle finder like you can buy at any lumber yard
or hardware store. At midday, when the sun casts a shadow directly north,
place one end of the board on top of the post or railing with the other end
resting at the end of the post's shadow (on flat ground, of course—otherwise

the angle won't be true). Place the angle finder flat on the top edge of the board and it will show you the sun's angle; turn it 90 degrees on its side and it will show you the optimum angle for your array.

Interestingly, if you take a measurement at midday on the spring or fall equinox you will find the sun's angle to be 90 degrees, minus latitude. At 40 degrees north latitude, for instance, the sun will be at 50 degrees on the equinoxes, and will change 23.5 degrees up or down as the seasons change.

Place the angle finders along a board at high noon to find the sun's angle.

Using this information, most people who do adjust their arrays do it four times a year, going from latitude plus 15 degrees in winter; to latitude in spring; to latitude minus 15 degrees in summer; and back to latitude in the fall. These seasonal adjustments give you the average optimal amount of sunlight for each period.

If you design your array with these three adjustment angles, you need to realize that the summer and winter angles will stay in place longer than the spring and fall angles. In other words, your summer and winter settings will last for about four months each; your spring and fall settings for about two months. If the sun angle is between two adjustments, opt for the lower setting; you'll capture more sunlight. If you want to improve on the four-times a year adjustment strategy, see the next section for an easy and incremental method of adjustment. (Or, if you are feeling rich, you can simply increase the size of your array by 15% and keep them in fixed mounts. Problem solved.)

Seasonal Tilt of the Solar Array

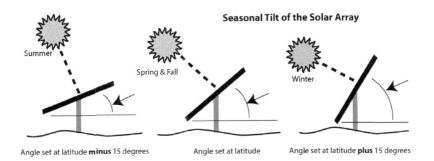

Angle set at latitude **minus** 15 degrees   Angle set at latitude   Angle set at latitude **plus** 15 degrees

Mounting the Modules

We began our solar venture with six 110-watt Solarex modules. Lacking the keen edge of experience, I built a massive frame to hold the 72 square feet of modules. The exterior frame and all the inside rails were made from 2 x 3/16-inch angle iron. It weighed about as much as a large man and it took four of us to carry it down the steep, rocky slope in front of the cabin to mount it on the concrete piers we had poured a week before. As unpleasant as that may have been, it was a picnic compared to the chore of carrying the frame back up the hill with considerably less help when we moved it to the log house.

Being several degrees wiser when it came time to mount the four 120-watt Kyocera modules we later added to the array (and another four, even later), I used 1½ x 1/8-inch angle iron for the outside frame, and 1½ x 1/8-inch strap iron for the inside rails. The finished frame weighed—and cost—about a third what the first frame did. It was easy to mount, and affords all the support the modules will ever need.

The third time around, when we decided to max out the system with the addition of four 175-watt modules, I built the frames from 1½- x 3/16-inch aluminum angle stock. It was expensive and difficult to weld but the frame is lightweight and professional looking, and didn't require a drop of paint.

While commercially manufactured, pole-mounted aluminum frames are available and widely used, if you're handy with a welder or an oxy/acetylene unit you may be money ahead by building them yourself. That way you can engineer them to fit the terrain and to adjust up and down the way you would like them to.

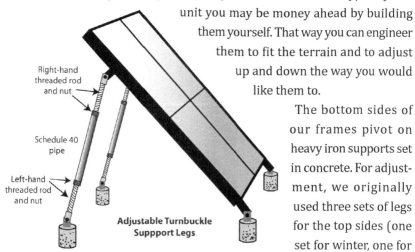

Right-hand threaded rod and nut

Schedule 40 pipe

Left-hand threaded rod and nut

Adjustable Turnbuckle Suppport Legs

The bottom sides of our frames pivot on heavy iron supports set in concrete. For adjustment, we originally used three sets of legs for the top sides (one set for winter, one for summer, and another for spring and fall). The only problem with this arrangement was that the modules in their frames are quite heavy, making seasonal adjustments a laborious two- or three-person operation. Since then, we have replaced the old rigid legs with two sets of adjustable turnbuckles I made from right- and left-handed one-inch threaded rod, with right- and left-handed nuts welded to either end of a length of sturdy pipe (the buckle). The first set of legs adjusts from 62 to 40 degrees, the second set from 40 to 20 degrees. The legs are easy to change out because the array does not have to be raised or lowered during the process. Best of all, I only have to give each buckle a few turns every so often to keep the array at the optimum angle, year-round.

Another way to minimize heavy lifting when changing the array angle is to buy (or design) a mount where the frames pivot in the middle. Centering the frames on one or two heavy steel posts is an ideal solution, providing you can set the posts deep enough in the ground—and with a broad enough concrete base—to prevent the whole array from toppling over in high winds. This solution wouldn't have worked for us; bedrock is less than a foot down where we set our array so we'd have been tempting fate by mounting the frames on central posts.

MICK'S MUSINGS

Mount your solar array on the ground so your poor sweat-gland-deprived dog can enjoy some great shade on hot summer days.

Trackers

Commercial mounts called trackers move the array in step with the sun as it travels from east to west across the horizon. They were a good investment when PV modules cost $10 a watt. Later, as PV became cheaper, the cost of a tracker could still pay off in the sun belt, but not in higher latitudes where, during the winter months, that the small amount of extra sunlight wouldn't be worth the extra expense or the hassle of having an added component that might occasionally require service or repair. Today, with the cost of solar panels becoming so affordable, it makes more sense to just buy extra modules with the money you would've spent on a tracker, since things that don't move are generally less of a headache than things that do.

Wiring the Array

In the dawn of residential solar-electric modules, all systems were off-grid and battery-based. There was no way to connect this energy to the grid until the mid-'90s, and no way to do it without batteries till almost 2000. Consequently, PV modules were designed as either nominal 12-volt or nominal 24-volt units with either 36 or 72 cells in series. Once conventional grid-tie became popular we started seeing a lot of 60-cell modules, which were easier and cheaper to make, and worked better size- and voltage-wise for grid-tie systems. With grid-tie dominating the PV market now, old-school 36 cell modules have all but disappeared, and 72-cells are getting rare. What's a battery-based fellow to do?

Fortunately, technology has saved us. All serious charge controllers for battery systems are MPPT types now (Maximum Power Point Tracking). Not only do these solid-state wonders extract 15%–30% more energy out of your PV array, they make it advantageous to wire common 60-cell modules in series strings of three modules each, and transmit from the PV array to the Power Center at 100 volts or higher. (See the explanation of series wiring below.) High voltage is your friend when it comes to pushing power around. So long as the PV array voltage is higher than the battery voltage, the MPPT charge controller can do its job, which is to run the PV modules at their maximum power-voltage point, and then down-convert that higher voltage into amps the battery can happily digest.

If you're running a conventional grid-tie inverter without batteries, installers typically wire the PV arrays in series strings of 11 to 14 modules (this varies a bit with particular modules, inverters, and climate), with string outputs of 300–550 volts feeding the inverter. This high DC voltage is one of the very good reasons do-it-yourself solar-system wiring parties are a bad idea.

In the old 12-volt systems, all the modules were wired in parallel—positive to positive, negative to negative. Thus, the amperage of each individual module adds to the total amperage of the array, without increasing the voltage.

In a 24-volt system, each pair of 12-volt modules was wired in series—positive to negative, and vice versa—doubling the voltage (from 12 to 24). Then each series string was wired in parallel to increase the amperage. Likewise, in 48-volt systems, sets of four modules are wired in series. The same holds for 60-cell (20-volt) modules: three modules per series string will yield 60 nominal volts, after which the series strings are connected in parallel to increase the amperage.

The important point to remember is this: parallel wiring always increases amperage while series wiring always increases voltage. To keep from getting these simple concepts mixed up, here's a good mnemonic trick: SERIES wiring results in SERIous voltage.

One other thing to consider when buying modules: only modules of identical wattage should be wired in a series string, since the amperage of the string will be equal to the amperage of the weakest module.

Basics of Wiring in Series & Parallel

| Two modules wired in **SERIES** **increases voltage** (48 volts at 10 amps) | Two **SERIES strings** wired in **PARALLEL** **increases amperage** (48 volts at 20 amps) |

WILLIE'S WARPED WITTICISMS
Dogs are like solar panels.
Heat 'em up and their
performance starts to ebb.

Fuses and Breakers

Beginning with the fragile aluminum channels on the surface of the individual solar cells and ending with the heavy copper leads going to the charge controller, the wiring of a solar array is like a complex watershed. Hundreds of tiny tributaries flow into dozens of larger ones, which in turn dump their current into a few even bigger channels, before emptying into the great river that ends at the load. (It doesn't exactly end at the load, since there is a pathway back to the source, but that's another story.)

As long as the integrity of the system is intact, the flow of current is orderly. But when a short circuit occurs somewhere along the line, large amounts of current can be sent back in a reverse flow—with potentially disastrous results. It's like what would happen if a cataclysmic seismic event sent the waters of the Gulf of Mexico raging back up the Mississippi River into each of its progressively narrower tributaries: the small streams would hardly be able to contain the additional volume of water.

To keep a similar occurrence from taking place within the confines of your array wiring, it's important to install fuses and/or circuit breakers wherever amperage is combined. Usually, the leads from each series of modules terminate into a combiner box, where the individual positive leads are directed through fuses or breakers of appropriate size. To reduce the risk of a costly mishap while working on the array wiring (and to satisfy the electrical inspector), the main lead coming from the combiner box is then run through a circuit breaker before going to the charge controller.

PHOTO: EVERGREEN

Grounding Your Solar Array

You should pound a ground rod into the earth next to the array and then run a heavy copper wire from it to the common house ground. Also, you will have to make certain that each frame, and each module within the frame, has a path to ground. You can use a multimeter to run a continuity test from the ground wire to the frame of every module, just to be sure.

Leaving Room to Grow

Even with the most assiduous planning, there is always a certain degree of guesswork involved in determining how many watts of solar power you will need. You don't want to be running the generator every time you wash a load of clothes, but you also don't want to ravage your kids' college fund to buy modules you don't need.

The best answer is to start with what you *know* you'll need, and then add more modules after you've had a taste of what the system is capable of. This is especially true if you are including a wind generator in the equation. Aside from ensuring that you have adequate space (on the ground, on the roof, or in front of the deck), all you'll really need to do when you set up the initial system is to run heavy enough wire to handle the amperage of the larg-

est system you would ever conceivably build, or run an extra conduit in the trench. It will save a lot of digging later. Then when you look at your array you can admire its inherent beauty, rather than thinking of all the work it might entail somewhere down the line.

If you are curious as to how solar cells really work, take a look at the appendix for an illuminating yet not too techy explanation.

– 8 –

Wind Turbines
Satisfying Benefits from an Invisible Annoyance

It's early April as I write these words. The mercury hangs just above freezing and the skies are thickly overcast. Beyond my office door a wind chime rings melodically. If I walk outside I can see our 1,000-watt wind turbine working, just beyond a copse of juniper trees. Though I can only feel a light breeze from where I stand, 50 feet above me the propeller blades are spinning furiously. I judge it to be about a 6- or 7-amp wind. The gentle whirring of the blades is a comforting sound. It's the sound of nature's energy being refined and transformed.

To me, it's music.

For the two years we were building our house, the wind sorely tested our sanity. Every time I tried to drop a plumb bob the wind came out of nowhere and deflected it. I'd throw a sheet of plywood over my shoulder on a perfectly calm day and instantly feel a playful breeze trying to wrest the sheet from my grasp. I'd walk up a ladder with it and the breeze would grow stiffer with every rung. If I moved the ladder to the other side of the house, the wind would change direction and follow. After awhile I became convinced that I could control the speed and direction of the wind, just by how I selected my activities. And who is to say I'm wrong? Not me; I'm a believer. But, even if a person can—by using plumb bobs and plywood—make the wind do a few parlor tricks, it's a fact of nature that you will never be able to outsmart it. Yelling at the wind or trying to ignore it just encourages it. Pleading with it will elicit no sympathy, whatsoever.

So you might as well face the facts: you'll never get along with the wind until you change the nature of your relationship.

Having put up with the wind through the entire construction of our house, I couldn't wait to hook up our new wind turbine and watch the wind do some good for a change. So, after laboriously erecting our tower and installing and wiring the turbine and charge controller, we had the pleasure of watching the propeller spin in the light breeze for about five minutes—before stopping altogether.

For two full days the blades didn't make a single revolution. By the end of the second day I began thinking that buying a wind turbine was the stupidest thing I'd ever done. What was I thinking? I could've bought 500 watts of easy-to-install solar modules for what I paid for one obstinate machine, sneering at me from 50 feet in the sky.

It was almost as if the wind wanted to know it would be appreciated before agreeing to do any work. And appreciate it I did, when it finally began to blow again after two days' absence. I believe it was the first time in my life I was actually happy to feel the wind. Obligingly, it blew more or less steady for the next several days, before settling back into its old, unfathomable non-patterns.

Since that rocky start two years ago, when I wasn't at all certain we'd made a good investment, I've begun to learn just how the wind factors into our energy equation. Though predicting the strength and duration of the wind at any given moment will never be more than a guess, over time the wind generally comes through for us.

Rex performs a routine inspection of his Bergey XL-1 turbine.

Would I now trade our 1,000-watt wind turbine for its monetary equivalent of solar modules? Not a chance. The wind provides us with power when the sun can't, which means less wear and tear on the batteries. Our wind turbine rarely supplies all the power we use during a stretch of cloudy weather, but it often provides enough to get us by. Then, when the sun finally does show itself—escorted by a breezy change in atmospheric pressure—the batteries recharge all that much faster.

Wind power is not for everyone, of course. In many areas of the United States, the force of the wind near ground level is too slight to be of practical value for a home-based wind system. And even if you do have usable wind power where you live, other factors need to be considered before you rush out and buy a wind system.

A Matter of Geography

In addition to the wind itself, you will need to have enough land to safely locate the tower and turbine. Usually, this means an acre or more. Some folks are able to avoid installing a tower by mounting one or more small wind turbines (500 watts or less) on a barn or other non-living outbuilding, where the vibration caused by the spinning blades will not shake the plaster off the walls and induce periodic fits of insanity. However, for larger machines (600 watts and up), a tower is a must.

There are two things to keep in mind when searching out a spot to place a tower. First, it should be far enough away from living areas and property lines that no one would be injured (or worse) if the tower fell over, or if all or part of the wind turbine came flying off in a killer wind or microburst—worst-case scenario stuff, in other words. How far is far enough? A good rule of thumb is 15 rotor diameters away from the house. So if your turbine has, say, an 11-foot propeller, you'll want to place your tower at least 165 feet away.

Secondly, the tower should be high enough to clear any obstacles that might be in the path of the wind. Ideally, the turbine should be mounted at least 30 feet above the tallest object—tree, building, hill, etc.—within 300 feet. This requirement is to minimize the effects of air turbulence, which is to a wind turbine what a washboard road is to a car. Besides causing extra wear and tear on the turbine, turbulence greatly diminishes the force of the wind.

How Much Wind is Enough?

If you have the land with an ideal location for a wind turbine and you've determined that neither your neighbors nor the local bureaucrats have any objections to a tower, you'll need to decide if you have enough wind at your site to justify the time and expense of installing a wind system. As you set out on this quest, the first thing you will discover is that exact wind data for your particular location is probably non-existent, unless your home is next to an airport or a military base. But you can still get a pretty good idea what the force of the wind is in your area by referring to the wind maps for your state. The Department of Energy (DOE) maps at the Bergey Windpower website (*www.bergey.com*) assign a wind class number to every square inch of every state. Though these maps are painted with a rather broad brush, they still offer a lot of insight for the wind resources that are available in your area. For further clarification, you should also read the National Renewable Energy Laboratory's Wind Energy Resource Atlas of the United States at *http:// rredc.nrel.gov/wind/pubs/atlas/*. This is a thorough document that discusses national and regional wind patterns, seasonal variations, and the painstaking methods used to compile the data.

A quick glance at a national wind map will show that the most paltry wind resources are in the Southeast, while large areas of excellent wind are in the upper Midwest, particularly the Dakotas and the western edge of Minnesota. Good winds can also be found in the higher terrain of both the Northeast and Northwest, and all along the Rocky Mountains.

You might also want to view the maps at the Southwest Windpower site *(www.windenergy.com)*. Southwest has collected links to the best maps available for each state. Some were compiled by NREL, others by TrueWind Solutions. In some states you will be able to input geographical coordinates and print out an extensive data sheet showing, among other things, the average strength of the wind from 16 different directions, as well as the average wind speed and power density during different seasons and from varying heights above the earth's surface. Bear in mind, however, that this information results from extrapolation of existing data from the nearest sites where measurements actually were taken. No one really knows for sure what the wind characteristics are at the top of the tower you haven't erected yet.

To be absolutely sure there's enough wind at your site you could buy an anemometer and monitor its readings for a few months. If you get a fancy recording anemometer, or one with a computer interface, you will actually be able to plot wind patterns over time to determine the average wind speed at your location for different months of the year. Should you go this route, however, please keep in mind that your anemometer readings will not be entirely accurate unless you are able to mount the instrument at the same height as your proposed wind turbine. The lower you place it, the less encouraging the results.

To make things easier, the folks at Iowa State University have compiled a Wind Energy Manual that will tell you, among other things, where best to locate your anemometer and how to extrapolate the data you collect to calculate probable wind speed at different heights above different types of terrain. Download this handy document at: *www.iowaenergycenter.org.*

But you really shouldn't have to sort through mountains of data, or spend a lot of money on wind monitoring equipment, for the simple fact is, if you think you have enough wind at your site, in all likelihood you do. As I mentioned earlier, my own personal rule of thumb goes as follows: *If the wind blows often enough and hard enough to annoy you, you can probably make good use of a wind turbine.*

However you go about determining if there is sufficient wind speed at your site, there are some surprising facts about wind speed and the amount of power you can hope to harvest from the wind that might be helpful to you. For starters, the relationship between the speed of the wind and the power it generates is not a simple linear correlation. What am I talking about? Just this: a 30-mph wind is not, as you might imagine, half-again as powerful as a 20-mph wind—it's nearly 3.4 times stronger! How can this be? It's because the force of the wind increases as the cube of the wind speed. So, 20 x 20 x 20 = 8,000, while 30 x 30 x 30 = 27,000. If you then multiply either of these products by 0.05472, you will discover the force of the wind in watts per square meter (W/m^2) at sea level for that particular wind speed. This is a tidy arrangement, because it turns out that solar radiation is also measured in W/m^2, so it's a simple matter to compare the speed of the wind hitting the blades of a turbine with the sunlight that falls on an array.

And how do they compare with one another? Generally, the power of the sunlight hitting the earth (or your solar array) in the middle part of a summer's day at mid-latitudes is equal to a steady wind speed of 22–23 mph—about 600 W/m^2.

This isn't the amount of power you'll be sending to your house, however. Your solar array will only be able to reap around 12%–15% of this energy, and these figures hold fairly well for wind turbines, too, though efficiency percentage is not a commonly used term with home-based wind turbines, owing to the fact that similarly rated machines may have vastly different sweep areas.

As you can see, it takes a pretty stiff breeze—square meter for square meter—to rival the power of the sun; far more wind than is blowing around in most locations. Considering the psychological effect wind has on a lot of people, this is probably a good thing. But it also makes your decision to install a wind system more difficult since you might live in a fringe area, where there may or may not be enough wind to make the installation of a tower and turbine a successful venture.

If the average annual wind speed where you live is 10 mph or more, you can almost be assured of having enough wind to reap a useful bounty of power from the unsettled atmosphere. This is because, unlike a solar array, a wind turbine's capacity to produce power is not limited to the hours between sunrise and sunset—it can produce power day or night, rain or shine.

The DOE maps list wind speed in power classes from 1 to 7. The upper limit of Class 1 winds approach 10 mph, provided the turbine is mounted high enough above ground. If you live in a Class 1 area you should probably do some homework before opting for a wind system. Living in a Class 2 area, though more promising than Class 1, does not assure you of enough wind, especially if your site is in a valley, near the lee side of a hill, or surrounded by towering trees (unless you are able to raise your tower at least 30 feet above the tallest nearby trees). By contrast, hilltops, coast lines and high plains make excellent sites for gathering wind. In particular, mountainous regions with large fluctuations in altitude are cauldrons of atmospheric change. Even without significant variations in atmospheric pressure (fronts), warm air will rise from the valleys during the day, while cool air will flow

MICK'S MUSINGS

A good wind is one that makes Newt's ears stand up straight. A great wind sends a cat or two spinning off the mountainside.

back into them at night, providing usable power while the sun is absent. If you are fortunate enough to have a home above a valley but below a distant ridge, you should have plenty of wind.

I cannot over-emphasize the importance of tower height in designing a wind system. There really is a lot more wind up there, and it will invariably be steadier and less turbulent than the gusty, chaotic breezes that occur closer to terra firma. In fact, the DOE generally considers the wind power density at 50 meters (164 feet) to be double what it is at 10 meters (33 feet). The actual figures will vary over different types of terrain, of course, but it's still an eye-opening exercise in mathematics. You probably won't erect a 164-foot tower, but the DOE figures do make a point: height is good (just like voltage).

Although most modern wind turbines begin to spin—and thereby produce some amount of power—at 6 to 7 mph (the "cut-in" wind speed), they will not really begin to produce much in the way of usable power below 8 or 9 mph. For battery-backup systems, this may enough wind to keep the batteries charged and help reduce your electric bill. If you instead opt for a direct grid-tied system, you'll want an annual average wind speed of at least 10 mph.

We never bothered to look at the DOE tables before buying a wind turbine because at the time we didn't even know such data existed. Instead, we made a judgment based on the annoyance factor and it paid off. You'll have to decide which criteria will work best for you, but the above mentioned resources should go a long way toward guiding you to an informed decision.

Wind Turbine Basics

Take a lengthwise coil of wire and rotate it between two oppositely charged magnetic poles and you will produce alternating current. Spin it at the right speed (frequency), and you'll produce a usable current for running AC motors and appliances. But try to charge a battery with this crude little alternator, and it won't work. Somewhere along the line, the AC has to be converted into DC.

Some wind turbines—mostly (but not exclusively) those that are rated at 500 watts or more—send 3-phase AC through the lines and into a charge controller, where a series of rectifiers convert it into DC for storage in the batteries. Once it senses that the batteries are charged, the charge controller shunts excess power to a heat sink.

Other wind turbines do the AC to DC conversion within the turbine itself, sending DC through the lines and into the batteries. Certain advanced models go a step further, incorporating electronics that sense the batteries' state of charge and adjust the rotor speed accordingly. On these models the propeller stops completely once the batteries become fully charged, eliminating the need for both a separate charge controller and a heat sink.

Is it better to send AC through the lines to your house than DC? Not really. The system voltage is more important than the type of current the turbine outputs. As we've already discussed, a wire that is rated to carry 40 amps of 12-volt current (480 watts) a distance of 50 feet, with 2% line loss, will carry 40 amps of 24-volt current (960 watts) 100 feet. In other words, by doubling the system voltage you can move twice the wattage twice the distance through the same wire, giving you a four-fold return on your investment. Bigger turbines, then, that need to be placed farther from the house, should operate at higher voltages. This means, of course, that your entire wind/PV system will have to operate at that same voltage. So if you plan to do wind, plan early.

While machines made for battery-based systems will generally be configured at the factory for 12-, 24- or 48-volt operation, turbines designed for direct grid-tied operation churn out up to 600 volts DC from the turbine which is converted to grid-compatible AC within the inverter. For this reason, direct grid-tied wind systems are more efficient than battery-based systems—up to 94% efficient—since there is no energetically expensive transformation from DC electrical energy to chemical energy within a battery and back again into electrical energy that must be further converted from DC to AC by the inverter.

Which system is right for you? What's true for solar is true for wind. If you live in the backwoods of Minnesota where blizzards and ice storms can tear down power lines in the blink of an eye and outages last for days on end, then you'd best be looking for a warm, cozy place to keep your batteries. On the other hand, if you live on the outskirts of a town with reliable grid power, then it might make more sense to opt for the more efficient direct grid-tied system.

Component-wise, there is very little difference between solar and wind systems. Wind/battery systems are set up just like solar/ battery systems. In fact, many wind charge controllers have additional inputs for a solar array. And for direct grid-tied wind systems, the Windy Boy inverters from SMA

America have the same features as the Sunny Boy and other comparable grid-tied solar inverters.

The main difference between solar and wind systems—beyond the obvious, of course—is that wind systems require more thought and more homework, owing to the plethora of different turbines and towers to choose from, as well as the widely varying wind conditions from site to site. If you are considering wind as part of your renewable-energy system, you should talk directly with the various manufacturers, not just the guy who wants to install the type of wind turbine he just happens to sell.

WILLIE'S WARPED WITTICISMS
A smart cat will always stay downwind of the dog, no matter how unpleasant the smell.

Will Wind Power Work For You?

This is a tough question that requires several regrettably slippery variables to be considered before answering. Where we live in Colorado, the sun shines over 300 days per year. I estimate the average wind speed at our house to be between 10 and 11 mph. Our current PV system is rated at 2,420 watts, our wind turbine at 1,000. I have not yet metered the wind output, but I'd estimate that it is around 10% of our total energy production. Were we to purchase a different 1,000-watt machine designed to run optimally at lower wind speeds—a heavier turbine with a bigger sweep area—we could reap a lot more wind than we currently do. But any way we slice it or dice it, we're in a good wind area.

If you live in the Southwest where the sun shines high and often, wind will not be as accessible a resource as sunlight, and you'll need to do a bit of research before deciding if a wind system is worth the investment. But if you live in the Northern Great Plains where the wind is fairly constant and the sun is a fickle companion, then you will probably want wind to figure prominently into your energy scheme.

Does anyone near you have a reasonably new wind turbine? Knock on his door and ask him about it. No amount of hair-pulling, divination, blind-guessing or rough-calculating can outdo solid experience.

Cost is always a consideration. Bigger turbines are not that much more expensive than smaller ones, but there are greater peripheral costs involved. Bigger means farther (from the house), higher (off the ground), and sturdier. While some small turbines can be mounted on a roof (provided the roof is built with standard trusses for mounting the mast), large turbines need to be mounted atop towers.

As I mentioned earlier, you can sidestep the tower issue completely by mounting two or more smaller turbines along the peak of the roof or next to the wall of a barn or workshop. You'll pay more per watt for the these smaller turbines, but make up the difference with what you'll save on the tower. The problem with this solution is that the turbines will be too low to the ground to take advantage of the really good winds, higher up.

As with any major purchase, you'll want to buy your turbine from a reputable manufacturer that stands by its products. Shop around and ask lots of pointed questions. Any salesperson worth his salt should be able to answer all of your questions about the various models and help you make an informed decision. Besides the pros and cons of different turbines, you will also want to know all the warranty details and how difficult it is to obtain warranty service. Who do you call with questions about installation and wiring? And who do you call once the unit is installed? Can you talk to a human without running through a computerized maze? It's good to know these things before you buy. After deciding that you can afford both the turbine and the tower to mount it on—and the extra wire—you'll need to consider how much work will be involved in getting the tower from the ground to the sky. Is it something you can do yourself with careful planning, or will there be extra costs involved in erecting the tower?

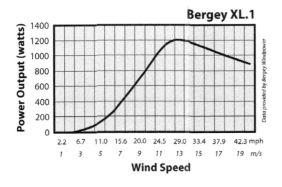

If all this is beginning to sound like a lot of work, that's probably because it is. But just think of all the work you'll be getting out of your turbine, once it's up and running. In the long run it will all be worth it.

Turbines: A Quick Look at the Windy Beasts

When home-based wind turbines are discussed, the image conjured up in your mind is probably that of a horizontal-axis machine. Consisting of a propeller, a rotor, an alternator and usually a tail, these turbines resemble wingless aircraft with oversized propellers. Though they may all look somewhat similar from a distance, there's a lot of difference between turbines, and your success or failure as a wind farmer will largely depend on which one you choose. Different machines are designed for different types of wind. Generally, machines with large sweep areas, such as Southwest Windpower's Whisper 200, are engineered to operate optimally in lighter winds. Other machines, including the Whisper 100, have shorter propeller blades and are designed to take the punishment meted out at hilltop locations and during severe storms. Still other turbines, such as the machines produced by Bergey Windpower and Proven Energy, can endure some really nasty weather and still perform fairly well in light winds.

Comparing wind turbines apples-for-apples will take a little research. Different machines have different cut-in speeds and different rated wind speeds, which is the speed at which optimal performance is achieved—usually in the 22- to 29-mph range. The table in the appendix lists the rated wind speeds, along with other pertinent data, for a few popular machines. It should be noted that there are many well-built turbines out there that aren't listed in the table. I'm not playing favorites; I just wanted to give you a good cross-section for the purpose of comparison.

Practically all turbines on the market today are 3-blade machines. A 3-blade machine runs smoother than a 2-blade unit and will be a little more efficient at converting wind into watts. As a general rule, the blades on smaller or lighter-duty machines are made from poly-propylene, while those on heavier machines are epoxy-coated wood or fiberglass. If damaging winds sweep across your site from time to time, you should avoid plastic blades on turbines of 1,000 watts or more. Trust me; this is the voice of experience talking.

A braking mechanism is also a handy feature, especially if you live in an area with ferocious gusts that could possibly damage your blades, or where ice storms might cover them with a layer of hoarfrost that can throw the system out of balance. With a wind brake—either mechanical or dynamic (electrical)—you can stop the turbine from spinning and wait for the sun to melt the frost or ice.

While the initial shock to your pocketbook will obviously be greater for a larger turbine, the ratio of wattage gained for money spent will also be greater. A larger machine will also outlast one or more smaller ones, and a heavy, slow wind turbine will have a longer lifespan than a light, fast one. So if you're going the way of the wind, buy as much as you can afford.

Wind Speed Conversion Formula
meters per second (m/s) x 2.23 = miles per hour (mph)

Towers: Holding Your Turbine Up in the Breeze

Once you have a pretty good idea what size and type of wind turbine will fit your needs, you'll have to figure out how you're going to hold it up in the path of the wind. There are four basic types of towers used by most homeowners: guyed-pipe, guyed-lattice, free-standing lattice, and tubular monopole towers.

Pipe towers are the cheapest, easiest to set up, and probably the most widely used. Made from sections of standard, off-the-shelf galvanized steel tubing, they are sleek, slim, and as inconspicuous as a tower can be—which isn't very. They are hinged at the base and then erected with the turbine already installed, blades and all. The major drawback to a pipe tower is that you cannot climb it for periodic inspections; it must be lowered.

Guyed-lattice towers are like ham radio towers. They are three-sided and of uniform dimension from top to bottom—mine is around 18 inches per side—and, like pipe towers, must be supported by a series of guy wires. They can be assembled either vertically by sections, or on the ground on a hinged base and tilted-up into place.

Free-standing lattice towers are broad at the base and taper toward the top, much in the same elegant way as the Eiffel Tower. Though more expensive (and showy) than pipe or guyed-lattice towers, you won't have to worry about clothes-lining yourself on a guy wire whenever you walk near it. Like the guyed-lattice towers, free-standing towers can be built in place, or assembled flat and raised to their vertical position.

A fourth type of tower, currently being offered by some manufacturers for various machines, is the **free-standing tubular monopole tower**. These are like the tapered steel towers you see holding communication equipment, or lights high in the air above highway exit ramps. They're expensive and require a crane to erect, but they're solidly built and good-looking, and they take up very little ground space.

Most turbine manufacturers offer tower kits sized for each of their turbines, and those who don't sell the kits will make recommendations on which towers will work best with a particular turbine. Listen to these folks—they know what it takes to hold their machines up in the wind. The lateral thrust put on a turbine in a high wind is mind boggling, and nothing you want to experiment with. You wouldn't put a V-8 in a go-cart, would you? Same difference.

A free-standing lattice tower.
PHOTO: BERGEY WINDPOWER

But even with good engineering, your tower and its foundation, like any structure, may be subject to regulation by your local building department. This means that you will have to comply with whatever codes are in place, since unlike an unpermitted workshop tucked inconspicuously away in the trees, it's a bit difficult to erect a tower without anyone noticing.

In any event, there's no substitute for sound engineering, so unless you're an engineer and rigger by trade, you should seek professional assistance to ensure that all goes smoothly.

I personally prefer a lattice tower for the simple reason that it's easy to climb when I want to conduct periodic inspections. I can also climb the tower (using proper safety equipment, of course) to wipe ice from the propeller blades after a freezing rain, rather than waiting a day or more for the sun to come out and melt the ice away.

Our lattice tower is a 50-foot 1950s affair that stood un-guyed for 40 years with a radio antenna mounted on top before I took it apart and hauled it up the mountain. It now rests in a 3-cubic yard block of concrete, set in solid bedrock. Its nine ¼-inch wire-rope guys are likewise anchored in bedrock. I'm pretty sure it could ride out a hurricane with a Volkswagen parked on top, though I doubt I'll ever find out. As for how I raised it, you're probably better off not knowing.

Raising the Tower

Once you decide on a tower, you'll need to determine if it's going to be hinged or permanently affixed to its concrete base. We didn't bother to install our tower as a fold-over configuration for the simple fact that there was really no practical place to fold it to on our rocky mountaintop.

For most people with enough open, reasonably level ground, however, a fold-over tower is the practical way to raise and lower your wind turbine safely and easily. The concept is simple and with a little careful planning the process should proceed without much difficulty. The basic idea is to hinge the tower at the base and attach a gin pole to it at a 90-degree angle. The length of the gin pole should be around one-third the height of the tower. Otherwise, as a turbine designer once told me, "the forces really start to add up." Near the point where the end of the gin pole touches the ground, you'll need a deeply rooted concrete pad with a pulley or a clevis firmly anchored to it.

When the business end of the tower is lowered and supported a few feet

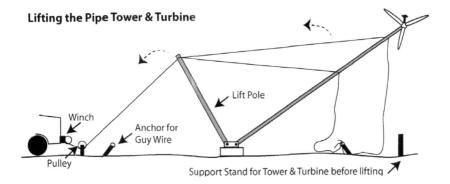

Lifting the Pipe Tower & Turbine

Lift Pole

Winch

Anchor for Guy Wire

Pulley

Support Stand for Tower & Turbine before lifting

off the ground, as it will be when it comes time to attach the turbine, the gin pole will be nearly vertical. The front-side guy wires should run from several points along the length of the tower to the tip of the gin pole, such that the full length of the tower and the end of the gin pole—all the places where stresses will be most critical during the lifting process—are supported.

On relatively light towers, a cable running from the top of the gin pole through a ground-mounted pulley is attached to a tractor or pickup locked into low range, with a steady-footed driver at the wheel, but you can just as easily use a bumper-mounted winch. In one precarious setup on a craggy mountainside we raised a 30-foot tower with a 3/8-inch battery-powered drill turning a small winch anchored into solid rock.

On heavier towers it is often desirable to run a ground-mounted cable through a pulley on the top of the gin pole then back to a winch firmly anchored in concrete. In this way the stresses are distributed across two cables rather than one, and the stationary winch allows for a slow and steady tower erection.

Guy Wires

However you raise the tower, it should be guyed to four deeply rooted concrete pads, with the two sets of guys perpendicular to the tower's movement set slightly back from the midpoint. This will allow extra support as the tower reaches the vertical position. The gin pole should also be firmly guyed on either side. These guy wires need to be anchored to the ground, in line with the pivot point of the tower, just as the two ends of a hinge are in line with the middle. Otherwise, the cables will bind as the tower is raised. Once the tower is in place, remove the gin-pole guy wires so the person who cooks your dinner doesn't trip over one and do a digger into the cold, hard ground.

Note: Wind tower grounding is covered in chapter 13.

Cost of Wind

Wind turbines are all over the map in terms of costs and quality. The same is true for the towers that hold them up in the air. For a good-quality commercially produced 1,000-watt turbine on a simple pipe tower, you should be able to get it up and running for as little as $5,000 to $8,000, depending on how much work you decide to do yourself. Build your own turbine in your spare time and you can shave a couple thousand off the bottom line (see below). Should you want a 2.5 to 5.0 kW turbine on a sleek monopole, plan on spending upward of $20,000.

Looking Deeper

In this chapter I have tried to address every aspect of wind power that I deemed necessary for you to consider before deciding what type of turbine to buy, and where—and on what—to mount it. At the very least, it should help steer you in the right direction. I know that LaVonne and I came to our decision with far less information than is presented here.

At the same time, I realize that the information contained here is, at best, a shallow scratch on the skin of a very deep body of knowledge. There are volumes of literature available on the subject of wind power. Hundreds of incredibly brainy people have devoted their lives to the study of harnessing the wind for the purpose of providing reliable, renewable, non-polluting energy. Yet, for all of that, it remains an elusive, empirical science. No matter how efficient an airfoil one person designs, someone else will always find a way to make it better. The same goes for turbines, and the associated electronics. There will forever be room for improvement, and that's the way it should be.

The wind is, after all, just so much air.

Electric Wind Turbines versus Water-Pumpers

Why aren't big, wide multiple blades used on turbines that make electricity? Water pumpers need lots of start-up torque to get things moving, which the large blade area provides. But once spinning, those multiple blades get in each other's airstream. Decades of careful experimentation have shown that two or three skinny blades will extract the maximum amount of energy from an airstream. Electric generators have almost no start-up resistance. Thus, the design of modern turbines with the minimum of skinny blades.

Do-It-Yourself Turbines

Can you make your own fully functional wind turbine? Yes, you can. But don't expect it to be easy or cheap. The Internet abounds with plans for turbines that will provide unlimited free power for very little time and investment. These claims are unadulterated hogwash. Most such plans involve attaching a rotor to a low-rpm DC motor and sticking it up into the breeze. Other plans offer variations on the ground-mounted vertical-axis design first patented by French engineer Georges Darrieus in 1931. Neither of these turbines can compare to a well-designed horizontal-axis machine. Attaching a propeller to a DC motor not specifically designed for the rigors of an atmosphere in turmoil is a disaster waiting to happen. And vertical-axis machines, no matter how well designed, suffer from a pair of insurmountable setbacks. First, they are mounted close to the ground where the wind is least active. And second, there is always as much mass pushing into the wind as pushing away from it, creating an unacceptable amount of drag.

So if you want to build your own turbine, plan on building a horizontal-axis machine that has been repeatedly proven effective over many years of service in all types of wind. The most promising DIY design I've seen, and the one I am most familiar with, is the axial-flux machine. Rather than the typical turbine geometry where a drum embedded with magnet spins around a central stator, the axial-flux design uses a pair of disc-shaped rotors than turn around a disc-shaped stator, rather like a pair of plates spinning in relations to third plate that sits stationary between them.

Commercially, the axial-flux design is employed by the Kestrel machines manufactured by Kestrel Renewable Energy, a subsidiary of Eveready, but plans and complete instructions for building your own are available. Where should you start? I suggest you buy a copy of *Homebrew Wind Power* by Dan Bartmann and Dan Fink, or *A Wind Energy Recipe Book* by wind-energy pioneer, Hugh Piggott. These guys spend their lives travelling the world, teaching their craft to a wide variety of folks. They know what they're talking about, as evidenced by the popularity of workshops. If you are serious about building your own machine, there is no better source.

- 9 -

Microhydro Power
Water + Gravity = Free Electricity

When I was a little boy I was fascinated by the giant water wheel my grandfather had set up in the irrigation ditch that ran next to my grandmother's garden. Round and round the wheel turned in the swift current, catching water in soup cans fastened to the paddles. Just past the high point, the cans emptied into a trough that routed the water into all the little channels that ran amongst my grandmother's carrots, radishes and peas. I would sit for hours in the branches of the apple tree that hung over the ditch, eating green apples until I was plump as tick and marveling at the industriousness of the ceaselessly turning paddle wheel.

It was my first encounter with hydro power.

My second encounter came years later, when I spent the summer with my brother on his cattle ranch in Costa Rica. There were two villages on the ranch, separated by three miles of jungle and pasture, and several hundred feet of elevation. The high village had no electric power other than that supplied by a giant diesel generator. Every night for three or four hours the generator would run so people could have electric lights for the time between sunset and bedtime, to read or sew, or do whatever they did under artificial light. The Ortiz family had the only TV in the village, and when the generator was running the local children would swarm to the big picture window in front of their house like so many moths to a yard light.

By a serendipitous circumstance of geography, the second village had no need of a big noisy diesel generator. Situated at the bottom of a hill 600 vertical feet below a fast moving stream, it had all the power it needed to supply the modest needs of a few primitive homes and a small commercial sawmill. A portion of the water from the stream above was diverted through a 12-inch pipe and sent shooting at incredible speed and pressure through a powerful turbine before emptying into the large river just below the sawmill.

I never knew all the particulars of that system—the output of the turbine's generator, the speed of the water, or how much water pressure developed in the pipe—because I didn't give a whit about renewable energy at the time. I did know that the power was always there for the taking, and it was free, though it was not as spellbinding as my grandmother's paddle wheel.

Microhydro power is a dream come true for those lucky enough to be able to use it. As marvelous as solar and wind technology may be, you are always at the whim of the weather. With hydro power, on the other hand, the energy you get when water flows from a high place to a lower place is fairly constant and, as long as there is water flowing, it takes the guesswork out your daily energy equations.

It's a little like the difference between day-trading in volatile stocks, or putting your money into high-yield bonds: with the former you may have a killer day (a steady 30-mph wind, with the sun beating down from an azure sky), but you could just as easily go bust (foggy and calm); with the latter (flowing water) you know exactly what you're getting day in and day out.

The question is, do you have the right currency to get into the game?

MICK'S MUSINGS

Solar panels need sun.
Wind turbines need moving air.
Hydro turbines need flowing water.
Dogs only need scheming cats.

Sizing-Up the Possibilities

If you want to create significant amounts of electricity, running water won't do you much good unless you are also able to use gravity to develop pressure. That's because the best way to create water pressure is to pour water on top of water. Commercial hydroelectric operations accomplish this by building giant dams across mighty rivers, then running the water through huge pipes (penstocks) into turbines located below. The sheer weight of all the water above and within the penstocks creates the pressure—0.434 psi per vertical foot of water—needed to spin the giant turbines.

Micro-hydroelectric systems differ from mega-hydro systems in that you don't have to create a monster lake out of your scenic, bubbling brook to make good use of the power. Instead, you simply collect a portion of your stream's water upstream, then route it downhill through a penstock (a pipe, generally in the range of 2 to 4 inches in diameter), where it naturally develops pressure. Once the water passes through the turbine it returns to the stream, unchanged except for the loss of a little kinetic energy.

In order to use microhydro power, then, you need to have a requisite volume of water running down a fairly good slope. The greater the slope the more power you'll be able to generate, since you will be able to achieve more "head" (the vertical drop from the water intake to the turbine) with a shorter penstock, and lose less power due to friction within the pipe.

Assuming that you have enough volume and head, all that remains to make this a perfect setting for a microhydro turbine-generator would be to have your house close to the stream. This is to ensure that there is little loss of power in the electrical lines running between the turbine and the batteries used to store the power.

But most of us are not quite so lucky. You're not likely to have a house right next to the stream, and the stream might not have the volume—or the drop—you'd like it to. So maybe you won't be able to consume free energy with utter abandon until the end of time. But you still might be able to get enough use out of your stream to augment a solar and/or wind system, and increase your comfort level proportionately.

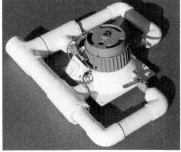

A pelton-type runner made of cast bronze, and a 4-nozzle, permanent magnet, generator-equipped turbine. PHOTOS: HARRIS HYDROELECTRIC

Microhydro Components

The Penstock

The penstock, or piping used to carry water to the turbine, can be made from a number of materials, including steel, plastic, or most commonly, PVC (polyvinyl chloride). It should run as straight as possible with few sharp bends.

You will need a strainer over the water intake to keep debris from entering the penstock and plugging the nozzles supplying water to the turbine. In cold-weather climates the pipe should either be buried or insulated to protect it from freezing.

The Turbine-Generator

The heart of any microhydro system is the turbine-generator. Similar to the turbine for a wind generating system, a microhydro turbine uses high-pressure jets of water to turn a propeller, also called a runner. The runner is attached to

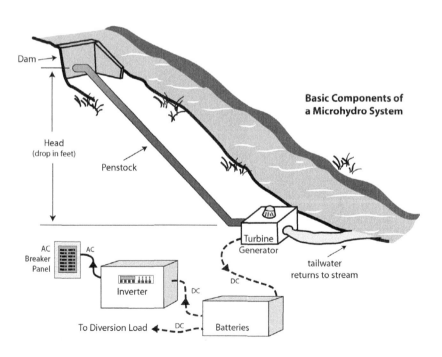

Basic Components of a Microhydro System

Dam

Head (drop in feet)

Penstock

Turbine Generator

tailwater returns to stream

AC Breaker Panel

AC

Inverter

DC

DC

To Diversion Load

DC

Batteries

a shaft running through the generator (or alternator) which spins a powerful magnetic rotor. The rotor induces an alternating electric current within the windings of the stator that surrounds it, which at some point is converted into DC current that can be stored in a battery bank. As with a wind generator, higher rpms produce more amperage, so you can think of volume and head of water as the equivalent of wind speed.

Most high-head microhydro systems employ a Pelton wheel for the turbine runner, which is a propeller with bucket-shaped depressions to catch the water. One to four high-pressure nozzles supply water to the runner. The number and size of nozzles used is determined by the volume of water coming through the penstock.

How much head do you need for this type of turbine? Depending on the volume of water making its way to the runner, you could begin producing usable power with as little as 20 feet, though more is certainly better.

I should also mention that since the generator attached to the turbine will probably not be designed to run underwater, you should plan to locate your turbine-generator along the side of the stream, above the flood plain.

The Regulator (Charge Controller) and Diversion Load

As with solar and wind power, a regulator (or charge controller) is needed to keep the batteries from becoming overcharged. A solar charge controller won't work, however, since these types of controllers simply disconnect from the power source once the batteries reach full capacity. This would leave the turbine to spin dangerously fast, causing a lot of expensive problems.

Instead, you will need a way to safely dump any excess amperage without leaving an open circuit. A diversion load, in other words. As with wind generators, the best way to bleed off excess power is to use it to produce heat, such as with a space heater or a water heater. In some systems the excess power is shunted to the heat sink before it reaches the batteries; in other systems the turbine's generator is hooked directly to the batteries, and any excess power is sent to the heat sink from the backside of the batteries through a diversion-load controller *(see chapter 10 on Charge Controllers)*. Again, the manufacturer or system designer will be able to offer various options, based on the specifics of your particular system.

Batteries and the Inverter

Once the electrical current passes the charge controller (and/or is shunted to the diversion load), your microhydro system will require no other special components. The current can be run into the same battery bank you use for your solar and wind, and can likewise be run through the same inverter. The only caveat to this would be if your hydro-generator was located a considerable distance from your house (as it might well be). In that case, the size and cost of the wire needed to carry the low-voltage current from the turbine to the rest of the system could become unwieldy *(see the appendix for determining wire size for various amperages and distances)*. If that's your situation, you might want to consider setting up a complete system (batteries, inverter and all) close to the stream, then running 120VAC to the house through smaller wires. Of course, this arrangement comes with its own set of problems, not the least of which is the extra cost involved in doubling-up on expensive components, in the event that you are also using solar and wind generating sources. Which way should you go? Unless you're a whiz at this sort of thing, this one is best left up to a qualified system designer.

Large Systems Without Batteries

If your proposed microhydro resources are sufficient to run your entire house—even during peak-load periods, when the well pump, air conditioner, clothes dryer and MIG welder are running at the same time—you should be able to install a turbine with an AC generator and avoid the installation of batteries and an inverter altogether. In these systems the AC generator acts as your own personal power plant, outputting 120 or 240 volts as either single or 3-phase alternating current. There is one crucial variable in direct AC systems that is not present in DC systems, namely the rotational speed of the turbine and, by extension, the generator driven by the turbine.

The frequency of your home's electrical current, as measured in cycles per second (hertz), needs to be held to a constant rate—60 hertz in the US, 50 in most of the rest of the world—for your appliances and other electrical devices to run properly. In generator-run systems, hertz is directly related to the rotational speed of the alternator producing the current, which in turn is determined in part by the electrical loads on the system; increasing the loads slows the rotational speed of the alternator and thereby lowers the hertz. Likewise, by lessening the loads you will increase the hertz. So ideally, you would want to continually draw just enough power from the system to keep it running at a constant rate, namely the rate that outputs AC at the desired frequency.

But this is a thoroughly impossible task. Any home's energy usage over time is a frenetic affair of ups and downs. It's never constant, not even at night; furnaces, boilers, fridges and well pumps kick on and off without you even being aware of it happening. So what's the trick? Simply this: you will need an electronic load governor working in concert with the generator, increasing the load—generally a heat sink of some kind—when house loads are low, and decreasing its activity when house loads are greater. The electronic load governor does this instantaneously and automatically every second of every day so you don't have to.

....

Direct Grid-Tie Your Microhydro?

Is it possible to tie a microhydro system directly into the power grid? Yes; in fact technically it's not all that difficult. You can either feed the generator's AC directly into the system, or rectify the AC into DC and use a direct grid-tie inverter to output grid-compatible electrical current. Since you will be tied directly into the grid, however, your system will be configured to go down every time the grid goes down, for the same reason a direct grid-tied solar system goes down; namely to protect the people who will be working to restore grid power from the lethal current produced by your powerful microhydro plant.

I should add a word of caution at this point, since you ought to be warned of the possibility of bureaucratic entanglements whenever you try to sell microhydro power to a public utility. This is because, unlike sunlight and wind, which no one owns, the water that flows in the stream in front of your house is someone's property and in all likelihood that someone is not you. So technically, by selling microhydro power back to the grid you will gaining monetarily from the use of another "entity's" property, and that entity is probably the state. Get the picture?

....

Getting Down to Particulars

If you suspect you might have a viable microhydro site, the next step is to take a few measurements and plug the numbers into the formula given at the end of this section. It will give you a rough idea of the amount of energy your site will produce. In particular, you will need to know the five variables for microhydro power:

1) **Length of pipe** from the water source to the turbine
2) **Feet of vertical drop** from the water source to the turbine
3) **Flow rate of your stream** (i.e., the portion delivered to the turbine)
4) **Length of the wire** (from the generator to the batteries)
5) **System voltage**

Rough Calculations of Power (using variables 1, 2 and 3)

The length of pipe and the feet of drop may take a little trial and error measuring. As I said before, the greater the drop the better. Even though there will be more resistance due to friction in a longer penstock, the extra pressure will more than make up for it.

Start with a likely spot upstream for collecting water in the penstock, then determine how much higher it is than the place where you wish to locate your turbine. This will give you the gross head. A quick and fairly accurate way to do this is to start from your turbine site by sighting down a handheld sight level, which is like a small telescope with crosshairs and a bubble level in the viewfinder. Train the level on a given place upstream—a tuft of grass, perhaps, or better yet an assistant's feet—then go stand in that spot, which you now know is the same height above your original position as your eye is from the ground. Then go to that spot and repeat the process. You can also work downstream using a tall grade stick held by an assistant. Each measurement you see on the grade stick minus the height of your eye will be the drop in elevation between those two points.

For a quick and dirty measurement you could try using an altimeter (such as the ones found on pricey watches, iPhones and GPS units) to get some idea of head, providing the atmospheric pressure does not change appreciably from one measurement to another. Reading the contour lines on a small-scale topological map will also yield a rough idea of how much head you'll have, although neither of these methods will yield an accurate enough measurement to base your decision on.

Once you know how long your penstock is going to be and what the drop in elevation is from the water source to the turbine, you need to know how much water you can collect in the penstock. Try building a small, temporary collection dam on the side of the stream and directing the water through a large pipe. Then time how long it takes to fill a 5-gallon bucket or some other large container. A couple of simple calculations will yield the gallons per minute. However, since our formula requires cubic feet per second instead of gallons per minute, we need to make one last conversion before applying the formula.

Let's say, for example, you've got a flow of 100 gallons per minute. To convert it to cubic feet per second, divide 100 gallons per minute (gpm) by 448.8, the number of gpm's it takes to equal one cubic foot per second (cfs): 100 gpm ÷ 448.8 = 0.223 cfs.

Finally, we're ready to run the formula. For the sake of our example, let's say that we have determined that we will have 50 feet of gross head, and we assume 55% efficiency. Our formula goes as follows:

Gross Head x Flow x System Efficiency x C = Power
Example: 50 ft. x 0.223 cfs x 0.55 x 0.085 = 0.52 kW (or 520 watts)

We see that our little system will produce somewhere on the order of 520 watts on a continual basis. This is equal to nearly 12.5 kWh's per day, which is plenty for many homes without wasteful appliances where the residents are mindful of their energy usage. LaVonne and I live quite comfortably, in fact, with two-thirds that amount from our wind and solar systems, and many of our friends get by on much less.

If you play with this formula a little you will make an interesting discovery—namely that 100 feet of gross head delivered at 50 gallons per minute is equal to 50 feet of head delivered at 100 gallons per minute. An exceptional flow, then, can compensate for low head.

This formula really gives just a ballpark number. You should use it to determine if it's worth the trouble to go to the next step. You will notice, for instance, that the length of the penstock is not entered into the equation directly; instead it is one of the factors you take into account when figuring the efficiency factor.

Refining the Numbers (using variables 4 and 5)

By this point you should know if you have a feasible microhydro site. The next step is to contact someone who can refine your numbers and make specific suggestions for the system components. Now you will need the last two variables listed earlier (4 and 5): the length of the wire running from the generator to the batteries and the system voltage.

This should be reasonably straightforward, especially if you already have a solar and/or wind system in place. If not, pick a likely spot for your batteries (inside a heated building, if possible) and measure the distance. If it's greater than 50 feet, you should consider setting up a 48-volt system to minimize line loss.

Once you have all these facts in hand, you'll need to find someone who can make sense of them. You could hire the services of an engineer, but you shouldn't have to. Most manufacturers of turbines will be happy to run the numbers for you. In any event, it never hurts to have more than one opinion (or one bid).

Power in the rough along the Big Thompson river.

If your numbers bear fruit and the cost of the system is not enough to make you blanch, then there is one more important step you need to take before ordering your components: you will have to talk to a county agent about the legal implications of your proposed plans. If the amount of water you intend to divert from the stream is but a fraction of the overall flow, then you will probably not encounter much resistance. But you still have to ask; water—and the natural habitats it supports—is a very touchy subject these days, and not one of those issues where forgiveness comes easier than permission.

Cost of Microhydro

A new Harris microhydro turbine runs in the neighborhood of $1,900–$2,300, depending on the number of nozzles. By the time you add in the batteries, charge controller and inverter you'll be looking at around $4,000 to $6,500 for a startup system, with additional costs incurred for piping and installation. You'll be better off in the long run if you can find a dealer who designs and sells complete systems.

Microhydro Terms

- **Gross head** is the actual distance of drop from the intake to the turbine (in feet or meters), not taking into account the friction developed in the penstock.
- **Flow** is measured in cubic feet per second (cfs), or cubic meters per second (m^3/s).
- **System efficiency** will be between 40% and 70%. It is affected by the length and condition of the penstock, the efficiency of the turbine, the gauge and length of the wires running from the turbine to the batteries, etc. If you're using a state-of-the-art turbine, have a relatively short distance from the turbine to the batteries, and do not have an inordinately lengthy penstock, start with 55%.
- **C (the constant)** is 0.085 when using feet; 9.81 when using meters.

Charge Controllers
Processing Your Batteries' Diet

The first charge controller I ever laid eyes on was a Trace C40, a nondescript white metal box with cooling fins on the top, about the size of VHS cartridge. I was not immediately impressed. But as soon as I began reading the accompanying manual my respect for the engineering packed inside the small unit grew from mild curiosity to utter fascination. It was, I quickly discovered, much more than the one-trick pony I imagined it to be.

The primary purpose of a charge controller is to charge the batteries without overcharging them. Different controllers have different ways of achieving this objective. Some work better than others. A charge controller will also disconnect the battery from the solar array after dark to keep current from flowing out of the batteries and back into the modules as they sit idle.

Outback's FM60 charge controller.

There are lots of inexpensive (under $50) charge controllers on the market that offer this bare-bones type of performance for small, unsophisticated systems, but for units capable of battery equalization or multi-step charging, be prepared to spend a bit more money. The more sophistication you can add to the processes of charging and equalizing, the better off you'll be.

Battery Charging

We installed the Trace C40 with a digital voltage meter in our guest cabin. It charges the batteries in three distinct stages: first, it allows the full charge from the PV array to reach the batteries until a preset voltage limit is reached. This period of unrestricted charging is called the bulk stage.

Once the bulk voltage setting is reached, the controller backs off the amount of current sent to the batteries in order to hold the voltage at the bulk setting for a cumulative period of one hour (the absorption stage).

After that, the controller enters the float stage, where the voltage is allowed to drop to a lower preset voltage where it will be maintained until the sun sets, or the (AC) loads exceed the DC input. The bulk and float voltage settings are determined by the installer and are set to voltages most practical for the specific application and the types of batteries used.

Why the complexity? Simply put, the various stages are needed to allow the batteries to "soak-up" a charge. If you use a multimeter to read the voltage of a battery as it's being quickly charged, and then disconnect the charger and take another reading, you will notice a significant voltage drop. If you let the battery sit for several minutes and take still another reading, you will see that the voltage has dropped even further. It may seem that the battery is mysteriously losing its charge, but it isn't; it's merely dispersing the charge throughout the cells. By charging the batteries in stages, the charge controller ensures that the batteries reach an actual full charge rather than an apparent full charge.

Equalization of Batteries

Equalization is the second important function of a charge controller, mainly for flooded, lead-acid batteries but also for sealed batteries at lower voltages. Sulfates can build up on the batteries' plates over time and affect their performance. If the sulfates crystallize on certain areas of the plates, those areas are no longer able to function. By bringing the batteries to a very high state of charge, most of the damaging sulfates dissolve back into solution, increasing the batteries' storage capacity.

The C40 equalizes the batteries by holding them at one volt above the bulk setting (2 volts for a 24-volt system, or 4 volts for a 48-volt system) for a cumulative period of 2 hours. The charge controller can be set to equalize automatically every 30 days, but that may be too often for a battery bank that is never deeply discharged. Besides, I prefer to initiate the process manually. That way I can pick a sunny, windy day when I know the batteries will equalize quickly.

Mismatched Voltages

Will a standard 12-volt charge controller work with a 60-cell (20 nominal volt) solar module? Sure, but you'll be losing a considerable amount of power in the process since any voltage above what the batteries will accept at any given time will be sloughed off and wasted. So if you're planning on using 60-cell modules, bite the bullet and buy an MPPT charge controller capable of down converting that extra voltage into usable amperage. You'll be glad you did.

Charge Controllers for the Wind Turbines

If you buy a wind turbine to supplement the output from your solar array, it should come complete with its very own charge controller. This is partly because every wind turbine has somewhat different charging characteristics, but also because wind turbines, being the feisty beasts that they are, require, shall we say, special handling.

Unlike the charge controller for your solar array which can simply step down the input as the batteries become charged, a wind charge controller must take more elaborate measures. This is because a wind turbine must be connected to a load at all times to prevent it from "freewheeling," which is to say, spinning out of control at dangerously high rpms.

The traditional means for doing this has been to run the excess wattage through a dump load, which is most often a separate component that heats either water or air. And while dump loads are still widely used, many modern wind charge controllers come equipped with sophisticated electronics that simulate heavy loads by shorting together the negative and positive leads.

In this way the charge controller can slow or even stop the blades as the batteries reach full charge, or the process can be initiated manually if you want to turn off the turbine for any reason.

Maximum Power Point Tracking (MPPT)

Several companies now manufacture charge controllers that maximize the charge by converting excess array voltage into usable amperage (power point tracking). This can be particularly helpful during the winter months when the modules are cold and therefore operating at higher voltages. Additionally, look for models that can use higher array voltage to power a lower voltage system. This can be very helpful for long wire runs if you're locked into a 24-volt system but your array has grown too large for the skimpy wire you have running to it.

Solar modules are designed to operate at higher voltages than the batteries they're asked to charge. There are two main reasons for this. Since voltage always flows from a higher potential to a lower one, the modules need to operate at a high enough voltage to charge the batteries in both low-light conditions (when the array voltage drops), and when the batteries reach a high state of charge. Since lead-acid batteries (in a nominal 12-volt system) often reach potentials as high as 15.5 volts during equalization, the array's rated voltage must be even higher. Typically, this will be in the range of 16.5 to 18 volts for a 12-volt module. During the bulk charging stage (the stage most charge controllers work in, most of the time), a typical charge controller will simply allow the full output of the array to run directly to the batteries. The batteries, of course, don't have any use for all that extra voltage, so they pull the array voltage down to a comfortable level.

"So what's the big deal?" I can hear you say. "If, as you've led us to believe, watts = volts x amps, why doesn't the amperage simply go up as the voltage drops?" Good question. The answer is it can—to a point. That point is the amperage the module was designed to produce. Think of it as a brick wall, 'cuz you can't get past it. At least not with a conventional charge controller.

Let's take a module from my array: a Kyocera KC120. It's rated at 120 watts, when it's producing 7.10 amps at 16.9 volts (7.10 x 16.9 = 119.99). How does this play out in the real world? Well, if the batteries are at 12.5

volts and they're drawing all 7.10 amps, the module is only producing 88.75 watts, or about 74% of its rated output. As the batteries reach a higher state of charge the percentage will go up, but there is always a significant percentage of power that is lost to heat.

An MPPT (Maximum Power Point Tracking) charge controller gets around this problem by using a DC to DC converter. It takes whatever voltage is optimal and converts it to the voltage the batteries are happy with. In the process, it uses the extra voltage to produce usable amperage in excess of what the modules were designed to produce. The trick is that the modules don't know this. They just think the batteries are finally getting their act together.

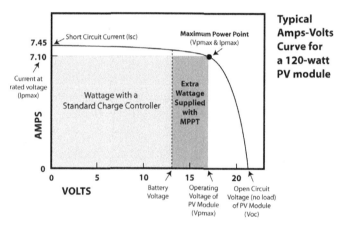

Power point tracking works best, then, when there is a large disparity between the PV module voltage and the battery voltage. This most often occurs when the batteries are heavily discharged, under a considerable load, or when the modules are cold. As the batteries reach a higher state of charge, loads are decreased, or sunlight heats up the modules, the extra power gained from MPPT diminishes. *(More on MPPT in the appendix.)*

....

The MX60 Charge Controller and the Cheap Freezer

LaVonne and I learned about MPPT charge controllers firsthand several years ago when we bought a little 5.5 cubic-foot freezer from a discount store, wondering if it'd be a ravenous energy pig. We were pleasantly surprised. At 400 watt-hours per day (according to our Watts Up? meter) in our 60-degree garage, it's nearly as watt-conscious as a high-efficiency model selling for seven times the price. But 400 watt hours is 400 watt hours, whether it's warm or cold, sunny or socked-in. It's more energy than it would take to run a 15-watt compact fluorescent day and night, or to pump 50 extra gallons of water per day from our 540-foot well. What were we thinking? Efficient or not, we knew our little freezer was going to take a painful bite out of our energy budget.

At first it wasn't too bad. Our customarily robust system handled the freezer well enough through the fall months, but once gloomy winter weather set in I found myself running the gas generator more and more often. My first temptation was to buy another 200 or 300 watts of solar modules to charge the batteries faster whenever the sun came out of hiding, but, as modules were above $4.00 per watt at the time, we decided to find a less expensive way to increase our daily harvest of precious watts.

That's when I looked at OutBack Power Systems' MX60 charge controller. If its MPPT technology really worked, it would save me the time and expense of expanding our array. The MX60's circuitry would also allow me to rewire the array for 48 volts—to minimize any line loss from the array to the controller—while still keeping the rest of the system at 24 volts.

The price tag on the MX60 was over three times what we paid for our old charge controller, but if it gave us an extra 75 to 100 watts throughout the sunny part of the day, it would be well worth the extra money.

Happily, the MX60 did the job and then some. Where before our array was maxed-out at 33 amps, I was now regularly seeing the amps in the 37–42 range, and even higher for short periods. The difference is especially noticeable on cold days when the sun shines through high clouds, or when direct sunlight hits the array after being obscured by heavy clouds. Once I wired the array for 48 volts, the output was even stronger than it was at 24 volts—about 5% overall.

We have since more than doubled the size of our solar array and have replaced the 60-amp MX60 with OutBack's 80-amp FLEXmax 80. We have also traded out our old propane fridge for a new electric model that requires nearly twice the wattage of our little chest freezer, so it is certainly gratifying to look at the charge controller on a cold winter's day and see that it is squeezing far more wattage out of the array than would ever be possible with a conventional charge controller.

....

Other Uses for Charge Controllers

Some charge controllers (Morningstar TriStar Series, for instance) have been designed to do more than regulate battery charging and equalization. Specifically, they may be used as either diversion load controllers, or as DC load controllers (though not at the same time as they are being used as charge controllers). Most of us have little need for either of these extra functions, but a short explanation of each may save you a moment or two of bafflement as you read the manufacturers' operating manuals.

Diversion Load Control

What's the purpose of a diversion load controller? To draw power away from a battery bank once it becomes charged. To illustrate, let's say you have a DC wind turbine or a microhydro generator wired directly to the batteries. In either instance it is easy to imagine how quickly the batteries could become severely overcharged and possibly ruined. You cannot just run these genera-tors through the solar charge controller and let it deal with the excess charge the same way it does with a solar array, because, as we discussed above, once a turbine is disconnected from a load the propeller—or impeller, as the case may be—will spin far faster than it was designed to spin.

By placing the charge-controller-turned-diversion-load-controller on the other side of the batteries from the wind generator input, however, and using it to divert excess current to a heat sink—a resistor that turns electricity into

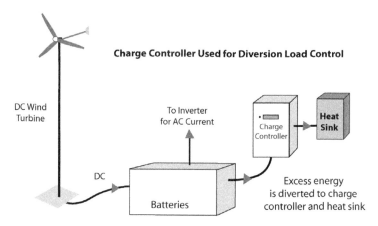

Charge Controller Used for Diversion Load Control

DC Wind Turbine

To Inverter for AC Current

Charge Controller

Heat Sink

DC

Batteries

Excess energy is diverted to charge controller and heat sink heat for heating anything from a room to a hot tub—the wind or microhydro generator remains connected to the batteries at all times without overcharging them, since any excess wattage sent to the battery bank will be harmlessly drawn off the backside.

DC Load Control

A DC load controller also protects your batteries, but rather than keeping them from becoming overcharged, a load controller prevents the batteries from becoming too deeply discharged. How? Let's say you have a DC refrigerator that runs directly from the batteries. What would happen if the batteries got too low to operate the compressor motor? It could damage the motor and possibly destroy the batteries. If it were an AC appliance, the inverter would simply disconnect itself from the load until the batteries were recharged sufficiently to once again supply ample current. But a DC load doesn't go through the inverter and so is afforded no such protection. Unless, of course, some other regulating component is installed between the DC load and the batteries. That's the purpose of a DC load controller: when the battery voltage falls below a preset level, the load controller disconnects the load until the batteries again reach a safe level of charge. If the low-voltage condition persists for an extended period of time—as could happen if your array was buried in snow while you were away from the house for a few days—it may cost you a lot of food and perhaps subject you to a memorable olfactory experience, but it's still a ton of money cheaper than a new high-efficiency refrigerator and a dozen or so batteries.

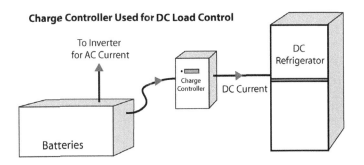

Charge Controller Used for DC Load Control

To Inverter for AC Current

Charge Controller

DC Current

DC Refrigerator

Batteries

Sizing for the Future

As with everything else, you should buy a charge controller that will be big enough to handle the extra amperage, should you decide to add more solar modules at a later date. As you almost certainly will. Conversely, you may design your solar setup in such a way that it can be split into two arrays, operating through a pair of charge controllers. This solution works well for large arrays and will leave you plenty of extra room to grow.

WILLIE'S WARPED WITTICISMS
The best way to keep a dog
from becoming overcharged
is to give its food to the coyotes.

Secondary Charge Controllers

A properly sized off-grid system will usually become fully charged by the middle of a sunny day. From that point on it shunts away more wattage than it delivers to the batteries or uses to run the normal house systems. In a word, it's wasted. So why not put that discarded energy to good use and have a little fun in the process by taking charge of the charge the charge controller eschews?

When we finally decided to get a "real" refrigerator—one that runs on electricity instead of propane—I never realized how useful it would be as a tool to regulate the batteries. Our standard Kenmore fridge was rated at 381 kWh per year, but with the help of our Watts Up? meter we quickly learned that with the ice maker turned off, the fridge uses only around 275 kWh per year, while it would use over 400 kWh per year if the ice maker ran 24/7, which it never would, since no one could ever use that much ice. So on days when the solar array is cranking out wattage faster than we can use it, we make ice. On cloudy days we shut off the ice maker and let the fridge coast at the lower energy usage.

The same goes for the AC pond pump we installed in our garden pond (aka, deer watering hole). Since it draws around 40 continuous watts it's hardly anything we'd be able to run all the time, so the pleasant fountain of water the pump shoots high into the air is a sunny-day extravagance. Consider it a reward for diligently watching the watts when it really counts.

~ // ~

Batteries
Getting to Know the Beasts That Hold Your Sunshine

Wind turbines whir, inverters hum, but batteries just grumble. It's only natural. Compared to the workout the batteries are subjected to every day, the other components in your PV/wind system have it pretty easy. Sure, the wind turbine is stuck up on a tower where it's beaten around by stray updrafts and errant cross currents all day and night, but it's an adrenaline junkie; to a wind turbine life is one big rodeo. The batteries know no such glory; they have to stay forever locked up in a dark box with no idea what's going to happen next. They alternate between being overstuffed with wattage to wondering where their next meal is coming from. They're in a perpetual state of uncertainty.

If anything in your system deserves a little pampering, it's the batteries.

The Right Batteries For The Job

What is the best thing you can do for your batteries? Never ask them to do more than they were designed to do. A battery is basically a factory where chemical transformations continually take place. Turn the key in your pickup's ignition and you will send hundreds of DC amps to the starter

Trojan's T-105 and L-16 batteries.

motor. Where do these amps come from? From electrons freed from bond-age as sulfuric acid is converted into lead sulfate, which then collects on the battery's plates. Once the motor is running, the alternator kicks in and repays the favor by sending DC amps back into the battery. The infusion of free electrons dissolves the lead sulfate and allows the free sulfur to revert back into sulfuric acid.

Not all batteries are the same, however. Car batteries are made with many thin plates with lots of surface area which facilitates fast charging and discharging, as when it is called upon to start the motor. These batteries are rarely discharged to more than 10% of capacity before they're quickly recharged by the alternator. A car battery has it pretty easy compared to what you're going to put your solar batteries through.

Only true, deep-cycle batteries will successfully perform the task of keep-ing the electrical systems in your home running smoothly day in and day out for years on end. The first time you lift a deep-cycle solar battery you will probably be surprised at how much more it weighs than a car battery of comparable size. Because they have fewer and thicker plates than auto-motive batteries, they are made to be charged slowly and don't mind being discharged fairly deeply.

A battery that falls somewhere in between a 6-volt deep-cycle battery and a 12-volt automotive battery is a 12-volt RV (or marine) deep-cycle bat-tery. You might think that you can use this type of battery and save yourself a few dollars and some confusing terminal connections, but you won't be happy when the batteries wear out in a couple of years. They also have thin-

---

Be Careful! Batteries are Dangerous! *Flooded lead-acid batteries vent hydrogen gas, which is highly flammable. As if that weren't enough, they are filled with a caustic brew known as sulfuric acid. Batteries need to live in a nice safe box with a sloped, hinged top; sloped so the hydrogen rises to the one-inch vent at the highest point (and so you won't pile stuff on top), and hinged for ease of access for maintenance. Don't allow sparks or flames anywhere near your batteries. And never, ever, allow a conductor (such as a wrench) to come in contact with terminals of opposite polarity. It could cause an explosion!*

ner plates than solar deep-cycle batteries, and are not made for the rigors of a PV/wind system.

Sealed, maintenance-free batteries (either AGM or gel type) are becoming more popular for PV/wind systems, largely because they can be stored in any position, never have to be vented, and never require the addition of water. You couldn't add water even if you wanted to, which means you need to be very careful when setting up the charging parameters; if they should ever accidentally become overcharged they would lose part of their water (in the form of hydrogen and oxygen gas) through the safety vent, and it couldn't be replaced. Moreover, batteries that have been deeply discharged and left in that condition for a period of time need to be deliberately overcharged to cook the lead sulfate from the plates, but it can't be done to any great extent when the batteries are sealed. So, while they may be less messy, sealed batteries do not allow any remedy should the plates ever become fouled. You'll be buying new batteries instead of restoring the ones you have.

This is not to say that sealed batteries should not be used, however. In fact, they are the battery of choice for grid-tie systems where the batteries are only called into action during those times when grid power is lost. Batteries in such systems have it pretty easy. They are rarely discharged, and any power that diffuses away is quickly replenished with a maintenance charge from the solar array or wind turbine.

Sealed batteries come in two basic types: **Absorbed Glass Mat (AGM)** and **gel-type**. AGM batteries use a liquid electrolyte that is suspended in a fiberglass material that surrounds the lead plates. Because the electrolyte is not free to migrate, it does not become stratified. Makers of AGM batteries include Concorde and MK Battery.

Gel-type batteries use a true jelled electrolyte, which serves the same purpose as the glass mats in AGM batteries. Because they have more moisture from the get-go, they can afford to lose a little through the built-in safety vents, though they should always be afforded the same care as AGM batteries. Manufacturers of gel-type batteries include MK Battery, Trojan and Hawker.

Amp hour per amp hour, sealed batteries will set you back more money than flooded lead-acid batteries, but they will last proportionately longer if they are properly maintained—neither overworked or overcharged—in a grid-tied setting.

In off-grid systems, on the other hand, the batteries are worked hard and will need to be fully charged and equalized often. For these systems, the only real choice are flooded lead-acid batteries. Probably the most popular battery ever to enter the solar market is the L-16 six-volt battery. Tall (about 17 inches) and heavy, it weighs in at around 120 pounds and holds around 400 amp hours of stored power. These batteries have a life expectancy of 6 to 7 years, but can last over 10 years if properly maintained and never discharged too deeply. Makers of L-16 batteries include Trojan, MK (Deka), and Rolls. Of these, Rolls batteries are the most robust, sport the longest warranty, and come with the highest price tag. I recently traded out a dozen worn-out Trojan's for a new set of Deka L-16s. They were just a little more than half the price of Trojans and the word is they last nearly as long. Time will tell.

For a lot less money you can set yourself up with golf-cart-style batteries. The Trojan T-105 battery is very popular in small PV systems. Providing 220 amp-hours of reserve power they have just a little more than half the capacity of an L-16. They're also cheaper; a T-105 battery will cost you about a third as much, but you'll be lucky to see more than five years of service out of it. We installed eight T-105s in the cabin in 1999. They later migrated to a toolshed before finally being recycled after eight glorious years of service. They owed their longevity to the fact that they were rarely discharged very deeply. By contrast, the 20 T-105 batteries we installed in the house in 2001 reached the end of their useful life after a little more than five years.

With the skyrocketing popularity of solar and wind renewable energy systems, your battery choices are growing practically by the day. Besides the tried-and-true 6-volt batteries mentioned above, you will find heavy-duty 4-volt, 8-volt, 12-volt and even 2-volt batteries suitable for use in off-grid and grid-tied systems. So look around; somewhere out there is the perfect battery for your application.

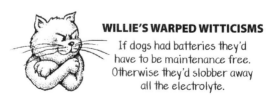

WILLIE'S WARPED WITTICISMS
If dogs had batteries they'd have to be maintenance free. Otherwise they'd slobber away all the electrolyte.

If You're Really Serious About Batteries....

You'll need a forklift to handle them, but once they're in place you shouldn't have to move HUP batteries for a decade or two. HUP stands for High Utilization Positive, meaning that the gradual corrosion of the positive plates—and the eventual demise of the battery—is greatly retarded by a special, patented process in which Teflon is incorporated into the positive plate material.

Solar-One® HUP batteries are sold in 12-volt units of six cells each, pre-assembled in heavy steel cases (euphemistically called trays). Power-wise, the nine different sizes range from 845 to 1,690 amp hours; weight per tray ranges from 642 up to 1,236 pounds. Big batteries, in other words. For system voltages above 12 volts, multiple trays are required.

These are flooded lead-acid batteries, which means they will require the same care and consideration you afford standard T-105 or L-16 batteries. On the upside, with fewer cells there will be fewer caps to fiddle with when it comes time to add water (and since they're all 25 inches high, you won't have to bend over quite so far to check them).

To set yourself up with these batteries, you should expect to pay some money; they cost about twice as much as a similarly rated bank of L-16s. But with a typical life expectancy of 15–20 years, they will assuredly pay for themselves over time. Just don't forget the forklift.

Solar-One HUP battery.
PHOTO: NORTHWEST ENERGY STORAGE

Pampering Your Batteries

Batteries are like draft horses: treat them well and they will reward you dutifully; treat them badly and they will tire out (or up and die), just when you need them most. And like horses, batteries don't require much to keep them happy; just a warm, dry place to rest, a little water now and then, and the security of knowing they will never go hungry. Nor does either object to being put to work, as long as they're not overworked.

Charging and Discharging

As I pointed out in the last chapter, a good charge controller will easily handle the chore of charging your batteries from wind and solar sources, as long as you give it plenty of amperage to work with. But when your batteries become greatly discharged after a few days of heavy loads and/or cloudy weather, you may need to charge them with a fossil-fuel-fired generator. In this case, the charging will probably be done through the inverter, not the charge controller. (See the next chapter on inverters for a full discussion of battery charging.)

How low can you let the batteries get before you need to drag out the generator? Lead-acid batteries can suffer permanent damage if they are ever discharged more than 70% of their capacity. This should never happen in a properly sized PV/wind system. Moreover, a well-calibrated inverter with built-in safeguards will shut down the AC loads before allowing the batteries to discharge to such a dangerous degree. (For large DC loads, a DC load controller should be used as an automatic disconnect. See the previous chapter for more information.)

To keep your batteries truly healthy—which translates to a long life expectancy—you should try to keep them above 80% of their rated capacity. The trick, of course, is in knowing when they have reached this level.

The easiest way to assess the batteries' state of charge is to install a meter that keeps track of the amp hours delivered to, and drawn from, the batteries. At a glance, you can tell how many amp hours below full capacity the batteries are, by subtracting the "amp hours from full" from the battery bank's rated capacity. Our meter, a TriMetric from Bogart Engineering, does the math for us, displaying a digital "fuel gauge" which shows the state of charge as a percentage.

I highly recommend a meter, but if you haven't yet installed one, the voltage readout on the charge controller can give you a fairly good indication of how the batteries are doing. If they remain 5% or more above their rated voltage under a heavy load, and 10 or more percent above under little or no load, they're doing fine. If the batteries fall below their rated voltage under a small load, however, they are in need of charging.

If you suspect a problem, check each individual battery with a multimeter. They should be the same within a few hundredths of a volt. A significantly low reading (around 4 volts for a 6-volt battery) could mean a dead cell.

At this point you should "weigh" the electrolyte by using a hydrometer to draw a sample of electrolyte from each individual cell in order to measure its specific gravity. In healthy batteries a reading of around 1.275 means the batteries are fully charged, 1.190 indicates a 50% charge, and a specific gravity of 1.155 tells you the batteries are dangerously low and in need of immediate recharging. (These readings are for batteries at 80°F; subtract 0.004 for every 10-degree drop in temperature.) Variation in either voltage or specific gravity indicates sulfation of the plates, meaning that it's time to equalize. An anomalously low specific gravity reading in any cell means the cell is dead and the battery should be replaced.

Equalization

Note: Deep equalization should only be performed on vented, liquid electrolyte batteries—the kind you add water to, in other words. If you have gel type and/ or sealed maintenance-free batteries, they can only be equalized at lower voltages. For these batteries you should follow the manufacturer's equalization instructions to the letter.

Equalization is really just a fancy term for the controlled process of overcharging your batteries. The purpose of equalization is to "cook" any sulfates from the plates that may have crystallized there, and also to "stir up" the electrolyte, which tends to become stratified over time. When the process is complete, your batteries should all be of an equal charge and hence "equalized."

How often should you equalize your batteries? Expert opinion varies from once a month, to once or twice a year. The Trojan Battery Company (who really should know, if anyone does) recommends equalizing only when low specific gravity is detected, or if the specific gravity varies widely (plus or minus .015) from cell to cell, and battery to battery. Others swear by a more frequent equalization schedule.

Most of us (myself, included) shy away from taking hydrometer readings at regular intervals. For us, "better safe than sorry" is the best rule of thumb. Batteries that are brought to a full state of charge often, and rarely, if ever, allowed to drop below 80% of capacity should easily be able to go two months without equalization. Batteries that lead a rougher life will need comparatively more attention. I adhere to a two-month schedule for our dual bank of 24 L-16 batteries.

As mentioned in the last chapter, a good charge controller will initiate and monitor the equalization process. It's all automatic and nothing that requires your attention. There is one detail, however, that I should point out, because it just might save you a lot of hair-pulling later.

The Trace (Xantrex) C-series charge controllers (among others) take the battery voltage 1 volt higher than the bulk voltage setting (2 volts on a 24-volt system, 4 volts on a 48-volt system), and keeps it there for a cumulative period of one or two hours. So if, for instance, the bulk voltage on a 24-volt system is set at 29.2 volts, the batteries will be brought to 31.2 volts during equalization. More sophisticated digital charge controllers allow a precise user setting, which is usually in this same range for most batteries. There is no problem with this, of course, unless the "high battery cut out" setting on the inverter is below the equalization voltage. If it is, the inverter will shut down once the voltage reaches the preset point, and leave you in the dark. I speak from experience.

Venting   *You do not need a fan to vent your battery box. Hydrogen gas is even lighter than helium, and can easily find its own way out providing, of course, the vent exits the box at the highest point.*

Remote Temperature Sensors

Wherever you keep your batteries, it is unlikely that they will stay a constant temperature as the seasons swing from cold to hot and back again. And even if they do, they will probably not be located in a room that's kept at 80°F. This is unfortunate because batteries require different charging parameters at different temperatures: the colder they get, the more charge they need to remain at optimum health. That's the purpose of a remote temperature sensor (RTS); it will keep your charge controller continually apprised of your batteries' temperature and adjust the bulk and float settings up or down as need be. The sensor has adhesive on one side and is thin enough to be wedged between two batteries. A cable attached to one end relays temperature information through a standard phone jack inside the charge controller.

Most inverters also have built-in jacks for RTSs. This is to accommodate those of you who routinely use a generator to charge your batteries through the inverter. If an RTS does not come with your charge controller or inverter, get one or two. They're about 20 bucks and change. Just be sure to get the right one; an RTS from manufacturer X will simply not work properly with charge controller Brand Y.

Don't Forget the Water!

Most batteries that land in the recycling heap before their time have simply died of thirst. And most of those batteries come from homes owned by people who have never used renewable energy before. That's unfortunate, because there is really very little work involved in keeping the batteries topped off.

The amount of water your batteries use will depend upon how new they are, how often they are brought to full charge, how hard you charge them, and how often they are equalized. Being maximally efficient, new batteries will require less water than older batteries. The same goes for batteries that are pampered compared to batteries that have it a bit rougher.

If you are just starting out, check the water levels at least once a month and immediately before and after every equalization. Keep records; write down when you water and equalize the batteries and how much water they take. If it's time to equalize and the batteries are a little low, that's okay; just make sure the plates are well covered but don't fill them up. If you do you'll create a mess and lose a lot of electrolyte in the process. After a few months you will develop a feel for when you'll need to add water. And after a couple of years a little built-in alarm will probably go off in your head long before the batteries reach a critical limit.

A procedure as simple as pouring a little water down a hole should not need much explanation, but a few pointers will get you started:

- **Use only distilled water.** Any other type of water will contain minerals that can dilute the electrolyte and collect on the plates, reducing a battery's effectiveness.
- **Bring the battery to a full charge before adding water.** Why? Because the electrolyte expands as the state of charge is increased. If you fill a battery with water and then charge it, acid will dribble out from under the caps. This creates a smelly, corrosive mess and also dilutes the electrolyte,

since some of it will have escaped. If the battery is greatly discharged and the plates are exposed, cover the plates with water before charging.

- **Don't overfill the battery.** Adding water to a level just below the bottom of the fill well is sufficient.
- **Never let the water level drop below the tops of the plates.** Exposed plates quickly begin to corrode. At a bare minimum, the plates should always be covered by at least ¼-inch of water.
- **Don't wear loose jewelry** that might cause a short between terminals of opposing polarity. It could be more excitement than you bargained for.

Some folks use a two-quart battery-watering jug for topping off the batteries, but with so many cables in the way it just didn't work for me. The method I finally settled on is about as easy as it can get. I mounted a one-gallon jug of distilled water about four feet above the battery bank. When it comes time to hydrate the batteries, I insert a length of small clear plastic tubing with a short length of ¼-inch soft latex tubing spliced to the other end. Then I create a siphon, which is to say that I draw from the plastic fitting on the terminal end like a soda straw. To control the water flow, I only need to pinch the latex hose as I move from cell to cell. This method gives me complete control.

The Battery Box and Keeping Your Batteries Warm

It is neither necessary nor desirable to store your batteries outside. As already mentioned, the optimum temperature for most batteries is 80°F. Efficiency falls off as they become colder, and outgassing increases as they become warmer. So, short of a climate-controlled room, the best place to keep your flooded lead-acid batteries is in a box within your house.

The box doesn't need to be fancy: I built both of mine from ½-inch CDX plywood. Many people use large plastic tool boxes. The only requirements are that the box be sealed, vented to the outside, and not placed under the inverter (so sayeth the National Electric Code). Door and window weather stripping works fine to seal

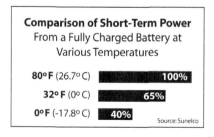

Comparison of Short-Term Power
From a Fully Charged Battery at Various Temperatures

80°F (26.7°C)	100%
32°F (0°C)	65%
0°F (-17.8°C)	40%

Source: Sunelco

the lid and one-inch PVC pipe at the highest point on the box is sufficient for the outside vent. A nylon screen covering or a plastic kitchen scrubber stuffed loosely inside the outdoor vent opening will keep insects and little furry varmints from exploring the inside of the vent pipe. If the box is going to be placed on a concrete floor, it's a good idea to support it with treated 2 x 4's to allow air circulation under the box and avoid rot.

Wiring the Batteries

Before you build the battery box, lay out different arrangements for your batteries on paper, because there is usually more than one way to connect the cables. The object of the puzzle is to keep all the cables—especially those going to the inverter—as short as possible.

The best way I've found is to lay out the batteries in rows, with each row being equal to the number of batteries in a series (a pair of 6-volt batteries for a 12-volt system, 4 for a 24-volt system, and 8 for a 48-volt system.) If they will fit the space you have allotted, it makes wiring a simple task.

Let's say for example you are running twelve 6-volt batteries in a 24-volt system. You will have 3 rows (series) of 4 batteries each. If you lay the batteries out in a 4 x 3 grid, each row of 4 batteries will be wired in series (positive to negative), to bring the voltage to 24 volts per row. After you do this with each row, you'll find all the remaining negative terminals will be on one end and all the free positive terminal on the other. These terminals will be used to make the parallel connections that combine the amperage from each series (row), without increasing the voltage. Connect all the positives together then all the negatives, then connect them to the inverter with heavy (4/0) cables.

Like water and most politicians, electrons take the path of least resistance. To make sure all of your batteries are worked equally, you'll need to make the path for all the electrons the same length as they travel from battery to battery and into the inverter. You do this by drawing the collective positive current from one corner of your battery bank and the negative current from the opposite, diagonal corner *(see illustration)*.

The batteries will require more care and maintenance than all the other components of your PV/wind system combined. Even so, they don't ask for much in comparison to what they give back. And after a few months you'll find that adding a little water and checking the connections now and then is no more of a hassle than taking out the garbage or shoveling snow off the deck. Or giving your wife a nightly foot rub.

Just think of your battery bank as your own private herd of short, compact Clydesdales. Batteries may not be as much fun to watch, but you'll get more work out them, and they're a whole lot easier to clean up after.

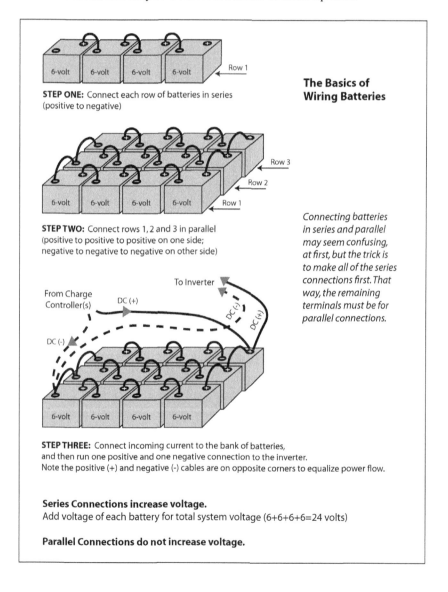

STEP ONE: Connect each row of batteries in series (positive to negative)

The Basics of Wiring Batteries

STEP TWO: Connect rows 1, 2 and 3 in parallel (positive to positive to positive on one side; negative to negative to negative on other side)

Connecting batteries in series and parallel may seem confusing, at first, but the trick is to make all of the series connections first. That way, the remaining terminals must be for parallel connections.

STEP THREE: Connect incoming current to the bank of batteries, and then run one positive and one negative connection to the inverter. Note the positive (+) and negative (-) cables are on opposite corners to equalize power flow.

Series Connections increase voltage.
Add voltage of each battery for total system voltage (6+6+6+6=24 volts)

Parallel Connections do not increase voltage.

Tip 1: Color-Code Your Battery Cables

Knowing positive from negative at a glance is helpful, safe and easy. Before wiring the batteries, lay out all the cables, separating the series cables from the parallel. Wrap a piece of red tape around one end of each of the series cables, and both ends of half the parallel cables. These will be your positive connections. It will help you avoid confusion later, and will probably impress the electrical inspector.

Tip 2: Connect the Positive Terminals Before the Negatives

If you already have enough spark in your life, connect the positive terminals before the negative when wiring a battery bank. Why? Because the negative leads offer a direct path to ground, a place the positive current really wants to go. By denying it the possibility of a quick and easy ride to the netherworld—like can easily happen when a breaker is still closed or a meter is searching for current, or when you accidentally touch both leads with a wrench or screwdriver—you will find your positive leads will be much better behaved. It won't make you immune to sparks (you can still short out both terminals on a battery, for instance), but it will improve your chances of an uneventful installation.

Tip 3: Battery Cable Connections

The National Electric Code (NEC) does not allow welding cable to be used for inverter cables and battery connections, although many backwoods folks still try to squeak by with it. In its place, THW cable (T = thermoplastic insulation; H = high temperature; W = moisture resistant) has become the cable of choice for battery connections. Most solar-supply houses sell it, either by the foot or prefabricated with terminal ends.

Home-Grown Hydrogen

With all the recent hype about biofuels, the versatility of hydrogen gas has been largely overlooked. But the fact is, homegrown hydrogen is eminently suited for off-grid systems. It's just a matter of waiting for the technology to catch up with the general principles involved. But someday soon it will be practical to create hydrogen at home using only sunlight and wind, and then use the hydrogen to run electrical appliances, heat the home, and maybe even power the family car.

As you know, current home-based off-grid renewable energy systems use solar-electric (photovoltaic) modules and small wind turbines to create DC electricity which is stored in large banks of batteries. When electricity is needed, the low-voltage DC travels from the batteries through a power inverter, where it magically becomes 120-volt AC house current. The inherently low efficiencies of PV modules and wind turbines notwithstanding, the solar/wind-to-electrical-load efficiencies for this type of system are fairly high; in the range of 80%–90%.

In a hydrogen-based system, by contrast, the batteries would be replaced by an electrolyzer, a hydrogen storage system, and a fuel cell stack. By going from electricity to hydrogen, then back to electricity, over 65% of the original

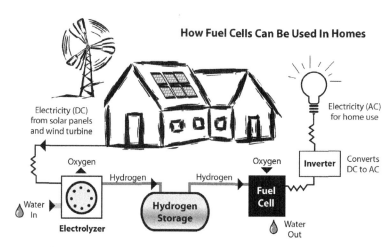

How Fuel Cells Can Be Used In Homes

Energy from the sun and wind can be used to create hydrogen by electrolysis. The hydrogen is stored until electricity is needed. A fuel cell then takes the stored hydrogen and adds oxygen from the air to create electricity. Heat created in the fuel cell can be captured for home heating. Water is the only byproduct.

energy is wasted using current technology, leaving you with a mere 35% of the power you started with. So why would you even bother?

A couple of reasons. First, since a fuel cell does not store power as chemical energy, it can deliver power faster than batteries, and will have a much longer useful life. Also, unlike a battery that generally keeps at least 50% of its power in reserve for optimal health, a fuel cell will deliver full power as long as there is hydrogen to fuel it, just as a gasoline motor will run full-bore as long as fuel is in the tank.

In addition to electricity, hydrogen can be used for home heating and cooking, and can even be used as car fuel. So how do you make the process less wasteful? You can begin by using a high-temperature fuel cell, such as a solid-oxide fuel cell. By doing this, the excess heat can be used for water or home heating, thus increasing the overall system efficiency.

Even if electrolyzers could someday reach 90% efficiency and fuel cells could have a combined heating and power efficiency of 70% or 80%, that's still less efficient than batteries. But what if PV modules were twice as efficient and half as costly as they are now? Then suddenly deluxe, home hydrogen-based systems could become even more affordable than today's battery-based system.

Teams of researchers all around the globe are working furiously to develop low-cost, high-efficiency solar cells. As solar energy grows in popularity, this research will only intensify. In the meantime we can look forward to advances in fuel cell, electrolyzer and hydrogen storage technologies. So, when it does all come together, it should happen fast.

To learn more about hydrogen technologies and the science behind it, read *Hydrogen—Hot Stuff, Cool Science (www.PixyJackPress.com)*.

....

— *12* —

Inverters
The Last Stop on the DC Trail

Before it gets to the inverter, the energy that will power your house goes through quite a ride. Energetic electrons knocked out of their orbits by particles of sunlight have charged through the wires at breakneck speed into your batteries, where they were pressed into service to convert lead sulfate into sulfuric acid. The sinusoidal positive/negative waves produced within the wind and microhydro generators were clipped and flipped and transformed into pulsing positive waves of direct current before joining the current from the solar array in the batteries' chemical energy storehouse. There is enough potential energy sitting in the batteries right now to run your house for several days, but until it goes through one final transformation inside the inverter, you can't even use it turn on one tiny light bulb.

The inverter is the magic box inside of which the DC harvested from the sun and wind and stored inside your batteries is finally converted into usable AC. As you might imagine, some inverters perform this task better than others.

The Magic Inside the Box

If Thomas Edison had gotten his way, no one today would be using AC and inverters would be no more than step-up DC converters. Fortunately, George Westinghouse (with more than a little help from electrical visionary Nikola Tesla), who espoused the virtues of alternating current, won the decade-long battle to provide lighting for New York City. The year was 1891. Since then AC has become the norm and sophisticated DC to AC inverters are a must for anyone wishing to use solar and wind energy to produce reliable grid-type power.

Alternating current is delivered in the form of a sine wave. This is a smoothly pulsing wave that gracefully arcs from a peak of positive voltage to an identical negative peak and back again. Essentially, the current reverses flow with each crest and trough. And it does it very quickly; in the United States, AC and the myriad things that run on it have been standardized to run at 60 hertz, or 60 positive-negative cycles per second. Most of the rest of the world pulses to a somewhat slower rate of 50 hertz.

As we saw in the chapter on wind turbines, a sine wave is the natural form an electrical current takes when it is produced by a coil of wire being rotated between oppositely charged magnetic poles. But how does an inverter, a solid-state device with no moving parts, take low-voltage DC—a flat boring stream of electrons—and teach it to do the high-voltage tango?

The heart of most inverters is the transformer. It turns low-voltage DC into the high-voltage AC we use to power our homes. Transformers, however, work on the principle of inductance, a phenomenon that only occurs to any significant degree in the presence of a pulsing (alternating) current. It takes some clever trickery to convince the transformer the direct current driving it is AC.

The magic behind the deception is a configuration called an H-bridge. Each of the two legs of the H have a transistor switch near each end—four switches in all—and the legs are joined by the transformer in the middle. The two bottom switches control the flow of negative current from the batteries, while the upper switches control positive flow back to the batteries. By electronically timing the opening and closing of the switches, the current

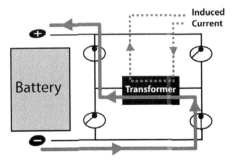

Induced Current

H-Bridge of an Inverter

By rapidly opening and closing switches on opposing corners, current flow is reversed and an alternating current is induced in the transformer.

can be made to flow first one way then the other through the transformer. Voilà—alternating current!

This basic configuration can be used to create modified sine waves, such as those produced by many inexpensive inverters. To produce a much better approximation of a sine wave (less than 5% harmonic distortion) a series of H-bridges and transformers of varying voltages are used, creating in essence a series of inverters whose outputs are mixed in ratios determined by the battery voltage.

Sine Wave Inverters

State-of-the-art sine wave inverters will produce cleaner, more precise current than the smoke-belching utility company on the far side of the mountain. Everything that runs on house current will run efficiently on the current they produce. As you might imagine, such technology doesn't come cheap.

We bought a Trace SW4024 sine wave inverter in 1999 when we first moved to the mountains. Initially, we installed it in the cabin, then moved it to the house once we had the rest of system installed. Besides one incident that wasn't really the inverter's fault (I'll explain what happened later) it has a nearly perfect track record. In its many years of service we've only seen two problems other than those resulting from my own programming errors: the sensitive electronics on our first propane range found the waveform intolerable. Also, after three years of service, the pump driving our solar hot-water system failed. Our consultant later informed us that, since our installation, he had learned that particular brand of pump (Wilo) did not perform well with our 13-year-old inverter.

Modified Sine Wave Inverters

For a lot less money, you can equip your off-grid house with a modified sine wave inverter, but I would advise against it. These inverters produce a stepped waveform that is really just a choppy approximation of a sine wave. For lights and toasters and vacuum cleaners, it may be good enough. But for certain other appliances, problems can arise.

After moving the sine wave inverter to the house, we bought a Trace (now Xantrex) DR 2524 modified sine wave inverter for the cabin. We were told by some people that it would eat an HP LaserJet computer printer in a hot minute. Others agreed, but said it would take a month or two to do it. I'm happy to say that we used that printer for years, even after running it for more than a year with power from our modified sine wave inverter, though there were times when the printer would stall for several moments before printing. The one appliance it would not run at all was LaVonne's Pfaff serger, which is a fancy sewing machine that cuts fabric as it stitches. As I later learned, the problem was not the machine itself but the rheostat that controls its speed.

The same thing happens when you try to operate things such as dimmer switches, variable-speed drills, battery chargers, or anything else where the current varies in intensity. All of these devices use solid-state switches called Silicon Controlled Rectifiers (SCRs) to control how much current is allowed through the circuit. To do this, an SCR needs a point of reference to continually reset its "clock," and the handiest one available is the instant when the slope of the sine wave passes through the zero-voltage point.

A modified sine wave, however, does not pass through the zero-voltage point at a gentle angle. It instead drops abruptly from 150 to zero volts, where it lingers for a moment before dropping abruptly again to the negative side of the waveform. With no distinct zero-voltage point, the SCR cannot effectively reset its clock. The SCR becomes confused and the tool or appliance will either not work at all or will behave erratically.

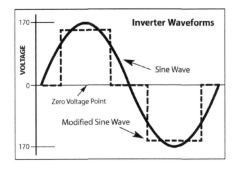

In addition, you may discover that fluorescent lights and stereo equipment produce an annoying buzz with a modified sine-wave inverter, and certain computers and peripherals that utilize SCRs may not work properly. Nor will electric clocks keep proper time, but that's really a blessing in disguise, since plug-in clocks are so wasteful you'll be doing yourself a favor by switching them out with battery-powered clocks.

In any case, at the back of the manuals for the old DR series inverters, Xantrex has compiled a list of devices that may experience problems running from a modified sine wave inverter. In addition to what I've already mentioned, the list includes microwave ovens, though the one we installed at the cabin works fine with the DR inverter. One lesson I quickly learned was that I could not let my DeWalt 12-volt drill batteries sit in the charger for very long after they were charged; otherwise they would quickly overheat.

The bottom line? If you are installing solar electricity in a weekend cabin, a modified sine wave inverter will probably work fine. But for a house with all the modern conveniences you have come to rely on, don't scrimp: a good sine wave inverter is the only way to go.

Inverter Functions (Battery-Based Systems)

What else can an inverter do, besides change low-voltage DC into usable AC house current? Plenty. A top-of-the-line inverter will have more features than you will ever use. Here are a few to look for:

High and low voltage shut-off is the inverter's way of protecting your appliances, the batteries and most importantly, itself. There should be one programmable setting for low voltage shut-off, and another for high voltage shut-off. The factory defaults are probably fine, unless (as previously mentioned) the high voltage shut-off is set lower than the voltage allowed by the

The power center of a large off-grid home (dual inverters in the middle plus two charge controllers in the upper right).

charge controller during equalization. If this happens, the inverter will turn itself off.

Battery Charging. You will also want an inverter that doubles as a battery charger. It should have settings for bulk and float voltages, and they should be set the same as on the charge controller.

Generator and Grid Tie-Ins (for systems with batteries). If you are completely off the grid, then the second feature (grid tie-in) will be of no interest to you. But if you are connected to the grid and still have batteries, a good sine wave inverter will stay in sync with the outside power source, cutting in when there is a power outage, or, depending on where you live, selling power back to the utility when there is an abundance.

Off-the-grid homes need a tie-in for a generator. Some inverters will start the generator for you—providing the generator is wired for remote start—when the batteries get too low, though you really should size your system so the batteries don't often reach that desperate state. Mostly, you'll use the generator after a few cloudy days when you want to run a heavy load, like a washing machine or a dishwasher. One thing to watch out for: make sure the "maximum charging amps" setting on the inverter is within a range that the generator can handle; otherwise it will trip the breaker on the generator. Even if it doesn't trip the breaker, the inverter might draw the generator voltage down below an acceptable level and disconnect itself. Something to keep in mind so you'll know what happened if it happens.

The search function is a feature we use at the cabin, but not at the house. It is designed to save power but it can cause problems. An inverter in search mode is at rest, but it sends out a pulse of current every second or so (this pulse-rate is adjustable) to see if anything gobbles it up. If it does—such as when you turn on a light—the inverter will come to life and power the load. This is all well and good until it finds a load that takes more than the preset search wattage to start but less than the search wattage to run.

Usually, this is just annoying. But if you are not careful, it can get expensive and even dangerous. For example, I was charging a DeWalt drill battery with the inverter set in search mode. (This is a different story than the previous one. Most of my inverter problems seem to revolve around drill batteries.) As long as the battery was charging there was no problem. However, after it finished charging and the inverter went back into search mode, the battery charger interpreted each pulse of current as a signal to start up again and sent a surge of power to the drill battery. Then, sensing that the battery was already charged, it would shut down, only to repeat the whole cycle. The result was a $50 battery bursting at the seams and too hot to handle.

Once we moved to the house, we simply left the inverter in the "on" mode. At absolute rest when the fridge and freezer are idle, our house consumes around 100 watts of power. The loads include: the inverter itself; three smoke detectors (required by the county to be hard-wired into the power supply); the displays on two charge controllers and two TriMetric meters; the clocks on the gas range and the microwave oven; and the on-demand water heater and the satellite box for cable TV. All other appliances that might draw a ghost current, such as the television and the computers, are plugged into surge protectors with disconnect switches.

Inverters and Power Panels. Most off-grid and grid-tie inverters can be "stacked" with one or more identical inverters. If your house has a number of heavy loads that might run simultaneously, or if you one day discover that the inverter you bought is too small for your growing needs, it may be wise to get two inverters and stack them, so that they operate in phase as a single inverter. The inverters may be wired in parallel, providing twice the amperage at the same voltage (120VAC), or they can be wired in series to double the voltage (240VAC). A transformer can be used to step the voltage up or down for certain loads, such as a 240-volt well pump. This being said, inverter manufacturers are coming out with ever bigger models to supply the demand for more powerful inverters (see below).

With renewable energy growing in popularity, people are building bigger houses with more appliances that can often tax the resources of a single inverter. For that reason, Xantrex, OutBack and others now manufacture power panels. These are pre-assembled units with one or two inverters and charge controller(s), along with a DC disconnect, transformer, and whatever else you may require. As you might imagine, power panels are expensive, so shop around.

OutBack Power Systems now has a number of inverters on the market, ranging in output from 2,000 to 8,000 watts, the largest being the new Radian series GS8048. OutBack pioneered the concept of an inverter module, which means that the programming is not done on the inverter itself, but from a

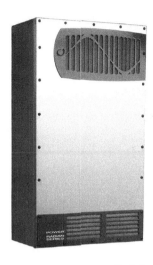

Outback Power's Radian Series inverter has a unique dual power module design that gives high efficiency operation while providing redundancy for critical operations and easy field servicing.

remote display unit called the Mate. Multiple inverters can be stacked without you having to shell out money for the display and programming electronics for each one, which is nice. And with all the programming being done on one remote unit, there's no chance of conflicts between different inverters. This is good, because no one wants their inverters locked in eternal battle with each other.

Xantrex offers a long line of inverters for all types of systems, including off-grid, gird-tie with batteries, and direct grid-tie. They also offer a complete line of small RV and marine inverters which often find their way into small, no-frills solar applications, such as weekend cabins and remote workshops. The highly successful SW-series has been replaced with new XW series of pure sine-wave inverters, comprising three models ranging from 4,000 to 6,000 watts. The DR-series inverters have been superseded by the TR series, which includes five models of modified-sine-wave inverters ranging from 1,500 to 3,600 watts.

Computer Interface. Most high-end inverters can be monitored and programmed via computer, as long as you have the interface and the software to run it. I haven't tried it myself, mostly because I really enjoy accidentally pushing the wrong button and shutting down the whole house once in a while. Still, I'm sure it would be handy to have sometimes; it's got to be better than resting on your knees on a concrete floor, pushing buttons with the hand that isn't holding the user's manual.

I know I've said it before, but I'm going to say it again: don't scrimp on the inverter! For a full-fledged house, a programmable sine-wave inverter is a must. Get a big one. That way, when the well pump is pumping water for the washing machine, you can still run your table saw without causing your own personal power outage. It will save you a few moments of button pushing, after several minutes fumbling around in the dark trying to find a flashlight. And it may spare you a derisive comment or two.

Direct Grid-Tie Inverters

The inverters used for direct grid intertie systems are a whole different breed of animal from the multi-purpose inverters used for battery-based systems. A direct-tie inverter wouldn't know how to charge a bank of batteries even if you wanted it to, and there's no place to hook up a gas generator. If the utility grid goes down, the inverter goes down with it, depriving you of any power you might be producing from your solar array and leaving you without backup power.

Direct-tie inverters are designed to do just one thing and they do it quite well. During the day when your solar array produces more wattage than your home is using, the inverter will power your home from the array and sell any excess back to the utility. At night when the array is idle, the inverter will prevent any grid power from trickling into the array.

While dispensing with all the circuitry devoted to battery charging and generator tie-ins, direct-tie inverters also do away with the need for separate charge controllers. Using maximum power point tracking (MPPT) technology, these inverters take direct current (DC) up to 1,000 volts, depending on the brand, and efficiently convert it to usable AC amperage. And, since they are designed to tie directly into the power grid, these inverters are generally made to be mounted outside the house.

Because battery-less systems are becoming so popular in certain places, there are a number of good direct-tie inverters on the market. Sunny Boy inverters from SMA America got into the game at the right time and have held onto a large market share ever since. SMA America offers a number of models from 700 watts on up, and can supply clean power for either single- or 3-phase applications.

For direct-tie wind systems, SMA has come out with the Windy Boy line of inverters. At present there are five models, inputting from 3,000 to 8,000 watts. Formerly sold only as a package deal with specific turbines, SMA now builds these units to be field programmable by certified personnel, thus greatly increasing the number of compatible turbines.

Other innovative direct-tie inverters that are daily proving themselves out in the field include Fronius, Power One, PV Powered, Solar Edge, Solectria, and Xantrex, which has made it back on the direct-tie scene with their GT-series inverters. You will find website listings for all these inverters in the appendix. Check 'em out. They each come with their own distinct features (too numerous to list here), so if you're in the market for a direct-tie inverter you shouldn't have much trouble finding one to fit your needs.

Micro-Inverters: Versatility in Small Packages

Sometimes it makes more sense to start small and work your way up; buy what you can afford and add on incrementally as your financial prospects brighten. It's an idea that works great with coin collections, herds of livestock, and balls of string, but not so great with bicycles, cosmetic surgery or hydroelectric dams. Or until fairly recently, direct grid-tied solar systems.

If, for instance, you begin with a 2,000-watt array and an inverter that maxes out at 2,500-watts of array input, you're limited to an extra 500 watts of PV. If you want to add more than that, you'll have to buy another (or a bigger) inverter, at considerable expense. Conversely, if you super-size the inverter from the outset and never increase the size of the solar array, you've wasted a lot of money on idle capacity. It really kind of means you need to know what you're going to do before you know what you're going to do.

Enter the micro-inverter, a hand-sized inverter designed to be mounted on the frame behind an individual solar module. Pioneered by Enphase Energy (and since embraced by other manufacturers), the micro-inverter takes a single module's DC output and converts it into grid-compatible AC, right on the spot. Because micro-inverters

Enphase
micro-inverter

allow for systems as small as 200 watts, anyone can get into the game, even apartment dwellers. As modules are added, they are effortlessly connected in parallel strings with plug-and-play cables (like lights on a Christmas tree) and ultimately wired into a branch circuit of your home's wiring. You will have much more leeway in the layout and placement of your array, since with high-voltage AC coursing through the wires, long wire runs are not a concern.

Nor is occasional or intermittent shading a problem. With a traditional inverter system, shading of one PV module—whether by windblown debris, nearby tree branches, chimneys, vent pipes, or snow—will cause all the other

modules in the series string to send a portion of their output to the affected module to try to "drive" it to a higher voltage. Not so with a micro-inverter system, since each module performs individually.

For information junkies, Enphase even includes an Energy Management Unit (EMU) that remains plugged into a wall outlet of you home. When con-

A grid-connected carport uses Enphase micro-inverters with each solar module.
PHOTO: MISSISSIPPI SOLAR

nected to the Internet, the EMU sends regular updates on each module's performance to a proprietary website that can be accessed from anywhere in the world. That way you can, from a beach in Tahiti, instantly surmise by a drop in performance that the neighbor kid's Frisbee is stranded on the third module from the east. Again.

For more detailed information on grid-tied systems, get a copy of *Got Sun? Go Solar,* a book I coauthored with Doug Pratt. A look at the advertisers in *Solar Today* or *Home Power* magazines will keep you abreast of new manufacturers and models.

WILLIE'S WARPED WITTICISMS
In cat lingo, an inverter
is anything capable of
flipping a dog over on its back.

– *13* –

Putting It All Together – Safely

Protecting You and Your System From Each Other, and Nature

I magine that lightning never wandered beneath the clouds. And while you're at it, imagine that you could be absolutely assured that every part of your system would work perfectly at all times—never requiring service or replacement parts or components—and that no one would ever try to draw more current through a wire than the wire could safely carry. If these three conditions were always true, then there would be no need for fuses, breakers, disconnects, ground wires or lightning surge protectors.

But this is planet Earth, we're all unavoidably human, and even when nature is on her best behavior, components still wear out and we still end up doing things with electricity that we shouldn't, though we claim to know better. Since this is the way of the world and there's nothing we can do about it, we—sentient creatures that we are—can (and most assuredly should) take precautions against the inevitable.

In this chapter I'll discuss grounding, over-current and lightning protection, and how important each procedure is to your home. I will then go through a simple off-grid system, beginning with the solar array, wind turbine and microhydro generator, and ending with the inverter. At each step along the way, I'll describe the minimum precautions you should take to protect your system and yourself from the forces of nature and any inherent dangers that might be present when replacing, repairing, or reworking any part of the system.

These recommendations are sound practices that have worked for me and other people I know with our particular systems—and the two or three by-the-book electrical inspectors that roam these parts. Your system (and maybe your inspector) may differ in subtle but significant ways, which means that it might require additional safeguards that are not mentioned here. As in all things electrical, the inspector has the final say. Nor is the National Electric Code a static thing. It grows, every year. What passes code today may not pass code tomorrow. What passes in one state or county may not pass in another. That's why electricians get the big bucks and the homeowner wires his house at his own peril.

It is not especially difficult to wire a basic renewable energy system, but it does take a clear logical mind, an appreciation for detail, and a lot of time. If you think you can put together your own system, you probably can. If you have any doubts, do yourself a favor and leave it to a professional.

This chapter is intended to be a cursory discussion of safety components and where they should be located in relation to the working parts of your system. Logistics, in other words. If you are hoping for specific information on wire sizes and types; conduit sizes and types; fuses, breakers, boxes or fittings, don't hold your breath, 'cuz you ain't gonna find it here. There's simply not enough room to cover that much ground in so few pages. Besides, it's boring stuff and every system is unique.

Grounding

The purpose of grounding a system is to provide an alternate path for current to flow should a current-carrying conductor (the positive lead from a solar array, for instance) ever come into contact with a metal surface (such as a fuse box) that you might touch. With a good unimpeded path to the earth where the charge can quickly dissipate, the current won't have to try and seek ground through you, because, unless you're wearing chain mail, you are not as good a conductor as copper wire.

On the AC side of your system, the (white) neutral wires, and the (green) ground wires will all ultimately lead to a copper ground rod outside the house. In turn, every light fixture, outlet box, fuse box, junction box, and PV/wind/hydro component encased in metal should have a path to ground. Any good electrician will know all about proper grounding of the AC side of things.

Grounding of the DC side is similar. All the negative leads from the wind and solar sources will ultimately be connected at the batteries, and all should have a path to the same ground rod as the AC side.

Bonding

The point where the AC neutral and ground wires join with the DC negative lead is called the point of bonding. This should be done at exactly one point within the system. It may be done at the ground rod or the inverter, but usually it is done at the AC service panel. If you install a bypass switch (to run your house from grid or generator power in the event your inverter fails) it can also be done there. All that matters is that it is done somewhere. Local codes may vary on the point of bonding. To be safe, explain to the electrical inspector that you wish to bond the DC negative to the AC neutral and ground leads, and then do whatever he or she suggests.

Lightning Protection

When preparing for the dangers of lightning, it's important to understand that lightning wants to go to the ground, so anything above ground is subject to its effects. Solar arrays are particularly vulnerable. During one especially worrisome storm at our cabin, lightning sent our charge controller into overload protection mode three times in less than a half hour. Fortunately, the only damage was a few frayed nerves.

Lightning occurs when there is a massive disparity in charge between the clouds overhead and the ground below. The idea of a lightning rod is to provide a path for positive charges from the ground to cancel out the negative charges in the air. In that way, a lightning rod helps to keep lightning from occurring in the first place, since it serves as a bridge between the sky and the ground where the needs of the former can be offered up by the latter. But if the clouds demand more than the ground is willing or able to give, then a lightning rod provides a path to ground that does not run through your house.

In many ways, solar arrays and wind towers act as lightning rods (so do stove pipes, for that matter). This means that they help equalize the electrical potential that exists between the ground and the sky. It also means that they will both attract lightning if the electrical potential becomes too great.

There is nothing you can do to change that fact. But there is a lot you can do to mitigate the effects of lightning.

Well Casings

Heavy metal well casings are notorious for attracting lightning. If lightning is a problem in your area, you may want to consider asking your pump installer to wire a lightning arrestor into the system to help prevent a damaging surge of electricity from zapping your inverter.

Log Homes versus Framed Homes

For those of you planning a log home, lightning protection is an absolute must in lightning-prone areas. This is one of the few things we found out the easy way. Being on top of a hill, we thought it would be a good idea to have lightning rods installed on our log house. Since lightning protection is something of a secretive art, we hired a professional installer to ensure that it was done right. When he saw that we had a log house, he told us how wise we were to install lightning rods, since log houses always sustain more damage from lightning hits than conventionally framed houses. Often, he said, a log house will burn to the ground from the excess heat it absorbs from a strike. Since logs are laid horizontally rather than vertically, a log wall presents a difficult path to ground. This means that lightning has to try harder. Bad news for the house.

The Solar Array

Each solar module is designed to carry only so much amperage. The bigger the module, the more amperage it can handle. When you wire modules in series you are increasing the voltage, not the amperage, so you can safely wire two or four modules in series without concern for overloading the module wiring.

But as soon as you wire one series of modules to another series of modules

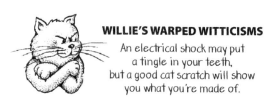

WILLIE'S WARPED WITTICISMS
An electrical shock may put
a tingle in your teeth,
but a good cat scratch will show
you what you're made of.

you are increasing the amperage and it is imperative that you isolate each individual series with **a fuse or a breaker** to protect it from a reverse electrical surge, as could happen with a short circuit somewhere in the line.

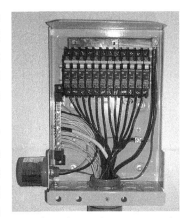

A combiner box with 12 circuits, and a lightning arrestor in the lower left corner.
PHOTO: DOUG PRATT

Typically, a combiner box is used for this purpose (OutBack Power and Midnite Solar both make good ones). The leads from a series of modules enter the box and run through a fuse or a breaker before its amperage is joined with the amperage from other series strings.

In addition, there should be one common disconnect for each array after all the current is flowing into a common feed. This is so the entire array can be shut down quickly and easily, either because of an emergency, or simply to service "downstream" components. A breaker box with a single, properly sized, DC-rated breaker works well for this purpose.

Grounding and Lightning Protection For Your Solar Array

Note: As of this writing the National Electric Code has no regulations requiring lightning protection for renewable energy systems. But if a lightning protection system is installed, it must meet their specific guidelines. Therefore you should consult with an expert if you plan to add a separate lightning protection system to your home or RE setup.

The array must be grounded. Period. Every module to every panel frame, every panel frame to the entire array, the array to a copper ground rod via a heavy copper ground wire. Obviously, the shorter the route to ground, the better. A buried ground rod beside the array and connected to the common house ground is ideal.

The grounding system and the system's fuses and/or breakers will protect your array from electrical accidents or oversights and most ambient electrical surges from nearby lightning. If you have a good charge controller, it will disconnect from the array even before the fuses and breakers can react. For more energetic strikes, a lightning arrestor is a very good idea. This is a small component that "absorbs" excess voltage, then slowly dissipates it. It should be mounted close to the array; either on the combiner box or on the main disconnect. Most have three leads: positive (red), negative (black), and ground (green). The ground wire may be connected to the disconnect box (if it's metal), and the box should have a path to the heavy copper ground wire running from the array to a common ground.

The Wind Turbine

Whether your wind turbine sends AC or DC to the house, the electrical inspector will want some sort of disconnect outside the house. This will be a switch that stops the propeller from turning (a **wind brake**), and may or may not be supplied with the generator package. If it isn't, you can order it separately.

Grounding and Lightning Protection for Wind

It may seem that a wind tower set in solid concrete and guyed to the ground with heavy steel cables would be well grounded. Unfortunately, it may not be. Since your wind tower will be the tallest thing around, you should take extra care to protect it from lightning.

To help avoid a lightning strike—and to minimize the effects, should one occur—drive a copper ground rod next to the tower and connect it to the tower with heavy copper ground wire. To be really safe, drive one ground rod next to the tower pad, and another rod at each of the guy wire pads. Connect all the ground rods together with heavy copper wire (#6 or bigger), connect them to the tower and each of the guy wires, and then run a ground wire to the common house ground. Bury all the wires at least six inches below the surface. Use heavy copper connectors, and make sure there is good contact (no rust, or paint). Expensive? In the grand scheme of things, not very. Worth it? If it's needed even once, yes.

A lightning arrestor should also be installed at the outside disconnect box, to help disperse any excess voltage that manages to get inside the lines. Since wiring differs from one type of wind generator to the next, you should ask the manufacturer what type of lightning arrestor to buy. Most likely, they will be happy to sell you one. They cost around 40 bucks.

Microhydro Safety

Your microhydro system should not require any special grounding or safety features beyond those already mentioned for solar and wind systems, though if (as is highly likely) the turbine is mounted on a metal frame it should be properly grounded.

Instead of an electrical or mechanical brake (as for a wind turbine), or a circuit breaker (like with a solar array), you will use the ball valves located near the nozzles to stop the turbine from spinning whenever you need to shut off the power. How well this flies with the electrical inspector (whose job it is to inspect electrical stuff rather than plumbing stuff) is anybody's guess. Just explain that (a) it's your only option; (b) it really does shut down the power; and (c) people use these systems everyday (even if your particular inspector has never seen one). Then hope for the best.

Charge Controllers

With outside disconnects for both the wind and solar charging sources, there need be no other breaks in the lines until the current moves past the charge controller(s), although it is certainly handy to have an indoor disconnect upstream from the charge controller. There should also be some type of over-current protection between the charge controller(s) and the battery bank. A properly sized inline fuse will work, though a DC-rated breaker will make it easier to isolate the battery bank from the DC sources and is more likely to bring a smile to the electrical inspector's face. (And we want the inspector to be happy, don't we?)

Some wind charge controllers come with their own built-in wind brakes. It is less common for solar charge controllers to have built-in disconnects. Plan accordingly. The breaker, or fuse, should be rated for slightly more amperage than the charge controller's maximum capacity.

DC Disconnect

The DC disconnect is usually a very large DC-rated breaker that lies between the battery bank and the inverter. It is designed to protect the batteries—and the inverter—should a short circuit occur somewhere within the battery bank. It also makes it very convenient for shutting down power to the inverter, and therefore the entire house, whenever you need to. The DC disconnect is sized for the inverter, and is designed to trip only when it senses far more amperage in the lines than the inverter could ever hope to use.

There are several companies offering DC disconnect breakers and boxes. Shop around. We use a Trace 250 amp DC disconnect; it has never been tripped. Because it's directly connected to the batteries, the DC disconnect is the favored place to connect all incoming and outgoing DC circuits, as for a DC fridge or ceiling fan, for instance. A good DC disconnect will provide mounting space for additional smaller breakers should you ever need them.

Safely Wiring the Components Together

With properly placed breakers and disconnects you will be able to wire most of your system without the threat of electric shock or a component-destroying short circuit. The solar array and the batteries are two exceptions. To avoid the possibility of damaging the solar array by accidentally crossing the incompatible wires on a sunny day, it's a good idea to cover the array with a blanket or tarp before trying to perform the delicate task of wiring the modules together in series and parallel. Or you can do your array wiring at dusk after the sun has set, though you may end up finishing the job with a headlamp.

The same goes with your charge controller. Many charge controllers can be damaged when live wires from your array are fed into it. So make sure you've thrown the breakers beforehand. Then when it's time to fire it up, turn on the battery power first to give its little brain a chance to boot up.

The batteries are another matter. By nature they are full of electrical potential and need to be wired with the utmost care. Know what you are going to do before you do it. Never work on the batteries without first isolating them from every other component within the system. This means flipping the disconnects (breakers) that lead from the charge controller and to the inverter.

There are basically two ways to coax a spark out of a battery: shorting together the positive and negative terminals on a single battery (wrenches and screwdrivers are the usual culprits), or shorting the positive and negative leads from a series or parallel string of batteries (usually this requires carelessness or confusion).

Know which cable is which! Color code all the cables with tape (positive = red; negative = black) before connecting them—either to the batteries or to the terminals of the charge controller, breaker boxes, or disconnects. Draw yourself a wiring diagram before you begin and consult it every step of the way. When finished, tape it inside the battery box for future reference. It may be that you know exactly what you are doing at the time you're doing it, but come back a few months later, and any wiring without color coding is going to look incomprehensible, especially on the AC side (where black = hot, white = neutral, and green = ground). At that point, you have to pull out the multimeter to decipher what you did.

A friendly word of advice for procrastinators: hide the multimeter under rolls of red, black, green and white tape. The tape will remind you why you are looking for the multimeter in the first place.

AC Wire Coding Colors	*Note: The NEC now states that black=positive and white=negative on DC systems, though in practice, this new coding is rarely used on DC-AC systems.*
Red or Black = hot	
White = Neutral	
Green = Ground	
	To avoid confusion, try using white for negative (DC) and neutral (AC) connections, and red for positive (DC) and hot (AC) connections.
DC Wire Coding Colors	
Positive = Red	
Negative = Black	

WILLIE'S WARPED WITTICISMS
Dogs would be far more manageable—but considerably less exciting—if they all came equipped with disconnects.

— 14 —

Backup Generators

Extra Power When You Need It Most

The last time we were in a major blackout Mother Nature dropped three
feet of snow on our high-plains horse ranch and the electricity was out
for five days. It was my job to figure out how to heat two mobile homes and
pump water for forty horses, all with my previously discussed 3.0 kW Coleman
generator. Fortunately, I had a roll of Romex on hand and was able to run
hastily wired isolated circuits from both furnaces and the well pump to the
generator outside. As if to accentuate the travails of a really trying week, the
generator's pull starter gave out on the fifth day. Two hours later the grid
power came back on. We were lucky.

Since then, we've moved to the mountains and off the grid and haven't
had to endure a power outage in over 13 years. It's one of the more satisfying
advantages of off-grid living: being perpetually prepared to keep the house
warm and the lights on. But not everyone plans to live off grid with a large
solar array and an all-business wind turbine tied into a huge bank of batteries
and a backup generator. You may instead opt for a direct-tie system in which
your present situation (like your solar array) is directly tied to the current
state of the electrical grid. Or you may still be weighing your options.

Either way, massive and lengthy blackouts are becoming more common and unless you live off the grid it's only a matter of time before Mother
Nature singles you out for a major blackout you can call your own. So the
question is: how can you protect yourself and your family from a possibly

life-threatening scenario without mortgaging your house or decimating your kids' college fund?

Generators: Choosing a Personal Power Plant

Being off-grid does not mean that you do not need a backup generator. With one exception, every off-grid family I know of relies on a gas generator to charge up their batteries when the clouds roll in for an extended stay and the winds are uncharacteristically quiescent.

Backup generators are powered by one of four fuels: gasoline, propane diesel...and wood gas, which will be discussed later in the book. Generally, propane and diesel generators are stationary (diesel because of their size, propane due to their necessary proximity to the fuel source), while gasoline generators can be moved from place to place. Many gasoline generators can be converted to propane (and back again), while diesel generators just run on diesel, or one of the bio-derived equivalent fuels available in some places.

The type of generator you choose will be determined by how much you intend to use it and what you plan to use it for. If you only run it once in a while, or like to move it from place to place (to do more than just charge your batteries or pump water from your well) you'd better stick with gasoline. If you can dedicate the generator to the status of backup-power-provider, then propane or diesel will be your best bet.

Buying a bigger generator than you need is wasteful and expensive. If you are living off-grid, size your generator to run your biggest conceivable load, such as your well pump, and be certain that you can pull that much amperage out of one circuit, because the biggest generator in the world won't do you any good if its power is dispersed between several small circuits.

Altitude matters. All generators are rated at sea level where there's lots of oxygen. As you leave the beach and move into the hills, the power will drop off at a rate of 3%–4% per 1,000 feet of altitude. It really begins to add up around 6,000 feet.

Our gasoline-powered Honda generator is wheeled out of the garage when needed.

Unless you're hard of hearing—or have a desire to be—buy a good, quiet generator.

While almost any generator with an electric starter can be rewired for remote start (via your inverter, for instance), if you think you will want this feature it will be cheaper and easier to buy a generator already wired for that purpose. Propane is your best choice for remote starting, since most gasoline generators have a manual choke and do not start easily if it's not engaged, and diesel generators are just plain hard to start in the winter, even with winterized fuel.

And finally, make sure you have a runtime hour-meter. If one doesn't come with your generator you can buy an add-on meter fairly cheaply. They are easy to install and will let you know when it's time to change the oil and perform other maintenance tasks. I keep a running oil-change schedule taped to the side of our generator. It's right next to the hour meter so I know exactly when I should change the oil and filter.

The same goes for those of you living on the grid who just want a reliable generator to plug your house into whenever the grid goes south. Like off-gridders, before you buy a generator you will first need to determine how big it should be. In your case, it should be big enough to power loads you'll absolutely need to keep running during a sustained blackout. Usually this is the furnace or boiler, the refrigerator and range, the well pump and a few lights and outlets. If all these things were running at once, how many kilowatts of electricity would they consume? Once you answer this question—either by reading each appliance's amperage ratings or, better yet, taking direct measurements with a watt meter—you can size your generator.

And like your off-grid neighbor, for the purpose of running your critical loads you will have to be certain that you can get all the power you need from one circuit. This will be 25 or 30 amps at 240 volts for the most popular 6- to 10-kW generators. Should you require more power, look for a stationary generator built specifically for home backup applications.

Not sure where to start looking? Honda offers a nice line of quiet, sturdy, all-purpose home generators. They're dependable

MICK'S MUSINGS

If humans knew the full extent of mischief simmering in their cats' convoluted minds, they'd force them all to run on treadmills and do away with gas generators.

machines that you can wheel out of your garage or toolshed whenever the lights go out, and you can easily connect them to the house via a heavy-duty extension cord.

If you'd rather go all out with a hard-wired stationary generator, take a look at the Kohler line. Kohler generators each come with their own all-weather enclosure and they're designed to be set on a concrete pad beside the house. And when wired into a Kohler transfer switch they can be programmed for automatic start when the grid fails and automatic shutdown the instant it comes back on.

There are many other manufacturers of dependable generators; I just happen to have had good experiences with Honda and Kohler. Whichever generator you choose, choose as if your life depends on it.

Transfer Switches: Where do I plug this thing in, anyway?

If you are at all familiar with your home's wiring, you've probably noticed that there isn't a neat little electrical box anywhere inside or outside your house with "generator inlet" or "auxiliary AC in" emblazoned on the cover. There's a very good reason for this: it is exceedingly dangerous to plug a generator directly into your home's grid-connected wiring. It's also illegal. Just the same, it hasn't stopped some gung-ho types from installing their own jury-rigged grid connections, often with disastrous results. Without safely isolating your generator's output from the power grid, you run several risks. When, for instance, the grid comes back on while your generator is hooked into it and running (and you know sooner or later it will), the grid current—which will be out of phase and flowing in the wrong direction—can (at the very least) burn up your generator's fragile windings.

But frying a generator is still preferable to turning it into a deadly weapon, a real possibility if you back-feed power into the grid when it's down. The high-voltage current from your generator will not stop at your house. It will instead follow every circuit available to it. You could easily electrocute a power company worker or even your neighbor, either of whom might be inadvertently exposed to hot wiring they assumed was not energized.

So install a transfer switch. Transfer switches are all designed with a single purpose: to make it impossible for the outputs of your generator and the power grid to ever run through the same circuits at the same time. Depending on the type and model, a transfer switch can be located either inside or outside your house and, as noted above, can be activated either manually or automatically. Many, but not all, have built-in critical-loads circuitry. If the switch you choose does not, you can isolate your critical loads from the main house panel into a separate subpanel which will then be routed through the transfer switch. That way (providing you've sized your generator to your load requirements) you won't run the risk of drawing more power than your generator can handily supply.

Conceptually, it's all pretty straightforward, but in reality the details get a little sticky. And when it comes the National Electrical Code, it's all details. Is this anything you can do yourself? Sure, if you've got the know-how to wire a house well enough to pass an electrical inspection. Otherwise, you'd be well advised to call an electrician. When it comes to electricity, nothing takes the place of experience. At the very least it may save you a few gray hairs.

Battery-Based Backup Power Systems

Some people take their backup power systems more seriously than others. Generally, it's a matter of geography: the farther out you live in the boondocks the longer it will be until the lights come back on following a blizzard, ice storm, hurricane or wildfire. Ironically, if you happen to live so far out that the power lines don't stretch all the way to your house, you're better prepared than anyone to deal with the circumstances that cause grid failures because you are, by definition, off-the-grid and not subject to the travails of power-grid calamities.

For you in the nether regions just a few miles short of off-grid territory, however, preparedness can mean the difference between life and death. A reliable, well-maintained generator is always your first line of defense, but even with a deluxe setup with automatic switching, startup and shutdown, any backup system that relies entirely on a generator has a built-in Achilles heel. A generator is going to use a baseline amount of fuel no matter how small of a load it's running, and the bigger the generator the more fuel it's going to burn. All day and all night.

What's the solution? Add a bank of batteries to your backup system to store up the extra energy your generator produces during the times it's running relatively light loads. In this way the batteries can supply the home's needs for hours at a time while the generator rests quiescently until called upon to refresh the batteries' charge. Plus, you will need a power inverter to run your critical house loads when the grid is down (as discussed in the previous chapter).

In the simplest system, a sufficiently large circuit from the main electrical panel supplies grid power to the inverter. The inverter, in turn, passes this power through to the critical-loads subpanel. When the grid is up and running, grid power is merely shunted through the inverter without being inverted, converted, subverted or perverted in any way. In fact, during normal times the inverter is really nothing more than a sophisticated battery charger; its only job is to occasionally redirect just enough grid power to keep the batteries topped off. Ho hum.

Once the grid goes south, however, the inverter becomes all business. In around twenty thousandths of a second the inverter kicks into action, drawing power from the batteries to run the critical loads. It's all very seamless and may in fact be so seamless you won't realize the grid is down until you go to turn on something not included in your critical loads, such as your 58-inch big-screen TV with Dolby surround sound.

··

The Sweet Smell of Living Off Grid

Some blackouts last longer than others, and if your power is shut down in midsummer and you have no backup system in place or are not around to take steps to avoid catastrophe, very bad things can happen. A recent wildfire in our area is a case in point.

For ourselves and the majority of people affected by the High Park wildfire of 2012, the mandatory evacuations lasted for nearly three weeks. During most of that time the electrical power servicing the homes in and near the fire perimeter was shut down for the safety of the firefighters who would be working near burned power poles and downed lines. It was an unavoidable action that made returning home a far from pleasant experience for most of those whose homes were spared. The source of the unpleasantness was the smell of rotting

A view from our backyard of the High Park Fire just a few hours after it was reported. Nineteen days and 87,000 burned acres later we were allowed back home. Thanks to our off-grid system, everything was just fine, including the contents of the fridge and freezer.

food, most notably decaying meat, from refrigerators and freezers left without power for weeks during which the outside temperatures routinely soared into the 100s. The solution was often more drastic than cleaning and scrubbing the fridge; after a certain time the smell leaches into the insulation beyond the plastic lining and it never comes out. Hundreds of evacuees had to replace their refrigerators and freezers. Others found that the unpleasant odors had escaped from their confines to permeate the very walls of their homes. One couple we know had put in a supply of crab legs shortly before the evacuation was declared. They found out the hard way that of all possible awful smells, rotting shellfish is at the top of the list. Not only did they have to discard their old fridge, their home had to undergo extensive ozone treatments and the sheetrock had to be sealed and painted.

Nor did the smell of rotting meat issuing from so many refrigerators go unnoticed by the hungry and displaced wildlife population. Throughout the 200-square-mile evacuation area homeowners were having to deal with black bears in their yards and, in some instances, inside their homes. So severe was the problem that bear-proof dumpsters for rotten food were brought in for the various neighborhoods.

LaVonne and I were exceedingly fortunate. We have often been thankful we live off the power grid, but never so thankful as the day we were finally allowed to return home. Being electrically self-sufficient, our house systems ran without a hitch during our prolonged absence. The outside light we turned on the night we evacuated greeted us as we drove up, and the house was just as we left it. It was, in every sense of the word, home sweet home. Minus the bears.

— 15 —

Getting Warm and Staying Cool
Options for Moving Heat

This section will be most helpful to those of you who are currently planning a new home powered by renewable energy sources. If you intend to
install a PV/wind/hydro system in an existing house, you won't have much
choice but to adapt it to the existing heating system. Fortunately, most heating
systems will work with renewable energy sources, though obviously some
are better than others. (On the other hand, if your current home uses electric
heat, you should either sell this book to someone else, or buy or build another
house. I strongly endorse the latter option.)

By the time LaVonne and I began planning our new log house, I had spent
seven years of my life living in a cabin with no heat other than a woodstove.
Before and after the cabin interlude—except for the years spent in my par-
ents' house, which was heated with hot water circulating through baseboard
registers—I lived in dwellings with forced-air heating systems. Between the
two, I far preferred the woodstove; it was quieter, easier to control, and cheaper
to operate. But after an absence of a day or two in the middle of January, it
took a steely constitution to come home in the dead of night.

Though LaVonne, poor girl, missed out on the earlier cabin years—chopping holes through the ice in the creek for bath water is one my fondest
memories—she was all too familiar with forced air heating. We were united
in our dislike of it. Owing to that fact, we were able to quickly focus our heating options when it came time to build the house we'd been dreaming about
since our courting days.

Obviously, electric heat was out; the PV/wind system needed to run it would rival the cost of the house. Likewise for a geothermal heat pump; super-efficient though it may be, it was, nonetheless, too energy intensive for the moderate off-grid installation we had envisioned. Propane wall heaters are a good, cheap, solution for cabins but quickly lose their practicality in large, multi-room houses. That left us with the hot water option. Baseboard or in-floor? It was a no-brainer: definitely in-floor.

The following chapters will cover geothermal cooling and heating, solar thermal heating, and biomass (wood) heating options, but let's first review the typical boilers and furnaces and basic backup options.

Propane Wall Heaters

Propane wall heaters are small, inexpensive units that can be mounted on a wall. They can be purchased either with a blower fan or without one. As backup heat for a cabin or small house with a woodstove, they're terrific. We have several friends who use them. The heaters save them the trouble of having to stoke the fire in the middle of the night, and keep their houses adequately warm if they're away for extended periods. We installed one in our (usually vacant) cabin a few years ago.

But a propane wall heater is really a brainless animal, with little hope of ever getting any smarter. It senses how cold it is in one location, then heats that location until its sensors conclude it's warm enough. It doesn't give a hoot how much heat makes it to the bathroom or the far bedroom, which is unfortunate, because a closed sleeping area is a bad place to install one. (With propane heaters there is always the risk of carbon monoxide poisoning. At the very least, you should have a good CO detector with fresh batteries at all times.)

So for a very small house on a no-frills budget, a propane wall heater may suffice. But for a large, comfortable house with multiple rooms, you will need something more.

Radiant Heating with Hot Water

Hot water heat has been around in one form or another much longer than natural gas has been available to fire the boilers that produce it. Even the

Romans used it to heat their floors. And today we still see cast iron radiators in many older homes, schools, and court houses. A lot of those old coal-, wood-, and oil-fired systems are still in use (now mostly converted to natural gas or propane), but their day is done; evolution has taken its course. Today's hot water heating systems come in two basic incarnations: baseboard registers and in-floor radiant heat.

Of the two, in-floor radiant heat provides the most even heat throughout the house. We chose it because it was the most logical choice for our home, for the simple fact that hand-hewn log walls—being irregular by nature—do not have nice flat baseboards to which you can attach the registers.

If you install radiant floor heat in your home—log, frame, straw bale, cinder block, adobe, beer can or waddle and daub—you probably won't be disappointed. This is not to say that it does not have its limitations.

The idea of radiant floor heat is to make the entire floor one huge wall-to-wall radiator. To accomplish this, a continuous length of extremely tough PEX (cross-linked polyethylene) tubing is snaked back and forth across the subfloor, stapled down, then embedded in a special type of lightweight gypsum concrete that is poured over the floor in a soupy slurry that hardens in two or three hours.

You can have as many or as few zones as you need, without the need of a nightmarishly complex system of ducts. Each zone runs on its own thermostat, which controls the pump for that particular zone. We have five zones in our house and garage. The open loft stays plenty warm from the heat rising from below.

Discovering the most efficient way to make use of our radiant floor heating system has been a real learning experience, owing to the fact that we heat the great room and kitchen with a centrally placed woodstove. Should we try to keep the thermostat in that zone set at constant temperature to pick up the slack at night when no one is awake to stoke the stove, or is it better just use one system or the other? We have concluded that (for that zone, at least) the latter solution is the most energy-efficient, even if we do sometimes awaken to a rather chilly house.

WILLIE'S WARPED WITTICISMS
The main reason humans warm their floors is to keep their annoying dogs happy. Cats just need warm humans.

Special Considerations for Radiant Floor Heat

An obvious limitation of radiant floor heat is the amount of time it takes to warm a room, once the thermostat is turned up. There is no instant heat as with forced air. This is because no heat can be felt until the hot water running through the PEX tubing raises the temperature of the medium in which the tubing is embedded. That's why radiant floor heat may not be the best backup heat source for areas heated primarily with a woodstove; by the time the floor is warm enough to pick up the nighttime slack you're already awake to stuff the stove.

Nor is radiant floor heat a good choice for drafty, poorly insulated, stick-built houses (of which there is no dearth in this country). But if you are building or planning to build a house using modern windows and highly insulating building materials, your home will produce and retain heat so quietly and efficiently that you'll have to trudge down to the mechanical room to even know when the heating system is working and when it is idle.

Log homes are especially suited to radiant floor heat because of the tremendous thermal mass of logs. When the house is warm the log walls soak up heat, then release it as the inside air temperature cools. In effect, the walls of a log house act as heat radiators in the same way the floor does.

You will be limited, however, in your choices for the finished floor. Carpet is feasible, though a standard, airy carpet pad won't work very well. You'll need a dense rubber carpet pad (not foam) to allow for the passage of heat. Considering the growing popularity of radiant floor heating, any reputable carpet distributor should know what will work and what won't. In my office, LaVonne put in FLOR carpet tiles, which aren't overly cushy, but they are fine when heating the room.

If you elect instead to go with a wood floor, it's best to install an engineered floor, one that "floats" on the surface of the gypsum concrete with no points of attachment. A floating floor can expand and contract without warping or buckling as it heats up and cools down. There are a few brands of engineered floors (not the thin, laminated variety) on the market; we went with one manufactured by Kahrs of Sweden. We're very pleased with it; it was prefinished and easy to install, and it's proving to be a good heat conductor (not to mention it stands up well to dogs' claws as they race for the door).

The third option, and probably the best, is tile or decorative concrete. Not only are they good conductors of heat, they do not need an insulating underlayment as is required for carpeting or wood flooring. We installed tile on our bathroom floors, wood in the kitchen and great room, and carpet in the office. Every room where the system is used stays plenty warm even on the coldest days, but the system does have to work a little harder to heat the office. Just the same, I enjoy the feel of a carpeted office, so—for me and the dogs at least—it's worth the small loss in efficiency.

Forced Air Heat

Most houses in America—both old and new—are heated with warm, forced air. The air is warmed in a furnace (usually gas-fired), then pushed through a labyrinthine system of ductwork with the aid of an electric blower (a squirrel-cage fan) into each room of the house through vents cut into the floor or ceiling. It's fairly inexpensive to install, but how efficient is it, compared to hot-water heating?

That depends on what is being used to heat the air that's being forced through the ducts. If you are using conventional fossil fuels such as natural gas, propane or fuel oil, then forced-air heat is less efficient than hot water heat, requiring more propane or natural gas to heat a house. This is mostly due to the physical fact that air is thinner than water. It goes places we don't want it to go, and requires more energy per unit mass to heat. On the other hand, if your heating system is driven by a geothermal heat pump, then forced air is the most practical way to go, owing to the fact that geothermal systems are by their nature able to heat air to higher temperatures than water (more details in the next chapter).

Many of the problems with forced-air heat are due to poor system design. Anyone who has ever lived in a house heated with forced air knows how easy it is to throw the whole system out of kilter just by shutting a door or two. Some rooms get too hot while others get colder. Often the only way to achieve any degree of equilibrium is to leave open all the doors to all the rooms—not always a satisfying solution.

This problem can be alleviated somewhat by providing properly placed cold air return ducts, which allow air to circulate throughout the house without building up pressure gradients in certain areas, pushing warm air out of the house in some places and drawing in cold air elsewhere. But as long as the entire house is run off a single, centrally located thermostat, the problem of uneven heating will persist.

Zoning is the obvious solution. By dividing the house into three or four distinct zones, each on its own thermostat, comfort levels can be maintained with very little effort (meaning you shouldn't have to constantly fuss with the floor vents). Zoning comes with a price, however. Besides upping the original cost of installation, each zone requires a small motor to operate a zone damper within the duct, in addition to the main blower motor located inside the furnace. It's not much, but for off-grid homes it can add up to a fair amount of wattage at the time of year you can least afford it.

That being said, great strides have been made in recent years to alleviate the inherent problems of forced-air heating systems. In addition to smarter duct system design, quiet super-efficient furnaces (with efficiencies now over 95%) are now able to incrementally adjust the airflow and fan speed, delivering exactly the right amount of heat when and where it is needed.

But perhaps one of the best things about forced-air heating systems is that you can have instant heat when you want it. This is particularly helpful if a large part of the heat for your home is provided by a wood stove. You can turn the thermostat(s) down in the evening and know that your heating system will kick in to quickly maintain the selected temperature(s), once the fire has grown cold—a feat a hot-water system cannot match.

Solar Air Panels

What is available for solar space heating? Solar air-heating panels mounted on the roof or onto the south side of the house is one solution that is gaining popularity. Similar to a flat-plate solar hot-water panel in size, appearance and function, a closed-loop solar air-heating panel draws air from inside the room into the lower part of the glazed panel and, with the aid of a fan, blows warm air back into the room after it has circulated through the absorber.

Many of these units use a low-wattage fan that is powered by its own small solar-electric panel, thus doing away with the need for electrical hookups. All that's required to install a solar air collector is to poke a couple of holes through the wall or roof for the ductwork. Then you're off and running. A wall-mounted digital thermostat tells the fan to shut off when the room is sufficiently heated. With a bit more work, individual or multiple units can be patched directly into your home's ductwork, pre-charging the furnace's cold-air-return with solar-heated air, thus reducing the amount of time the furnace has to run during each cycle.

Of course you might think a large, black panel attached to the side of your house is something of an eyesore, and if that's the case then a roof installation might make more sense for you. Wall-mount installations are fairly simple, so long as you don't have to tap into the home's wiring or ductwork; roof mounts require a bit more labor and expertise.

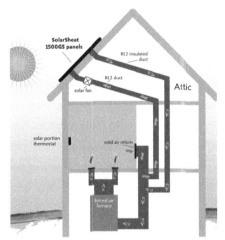
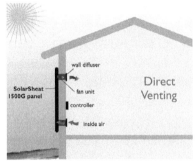

Large, multiple-panel systems can connect with the home's forced-air furnace. A simpler direct vent system installs on a south wall and has a built-in solar PV panel to power the fan.
COURTESY OF YOUR SOLAR HOME INC.

There are several U.S.- and Canadian-made models of solar air panels on the market. Before buying, compare output Btus and airflow rate, ask for customer referrals in your area, and check out the warranties. An SRCC certification is a good indicator the panel you are buying is a good one.

Expect to pay on the order of 15 to 25 cents per hourly Btu. The good news is that manufacturers estimate your investment can be recouped in four to eight years, depending on energy prices. With so few working parts, a solar air heating system will last a very long time.

– 16 –

Geothermal Heat Pumps
Home Heating and Cooling from the Good Earth

Most of the energy solutions in this book come from the great, untapped potential of the sunlight and breezes overhead. Mount a solar-electric array in the midday sun and you'll have ample electricity. Or stick a wind turbine high up on a tower and let an unsettled solar-heated atmosphere toast your bread and run your stereo system. For hot water all you need are a few solar hot-water collectors on your roof where the sun can do what the sun does best. Firewood, which is stored sunshine, is a pleasant source of heat for anyone with a chainsaw and a ready supply of dead trees. It's all quite elementary; the sky is where it's happening.

And yet for home heating and cooling, there is another source of energy—largely but not entirely originating from the sun, that may be the best solution of all. It lies right beneath your feet. Say 5 to 8 feet down. That's about the level where the effects of the different seasons are so gradual they barely fluctuate; where all the year's hot days mix with the cold ones in a terra-firma stew that never really ever warms or cools. In the northern states the deep ground will stay a fairly constant 45°–50°F all year long, while Southerners can expect constant ground temps of 50°–70°F.

But is this really good news for those of you living in the North Dakota hinterland, where winter heating bills rival the mortgage and cordwood procurement becomes a blinding obsession? Yes; especially you. It's just a matter of changing the way you think about heating.

In many ways the Earth is like a magic battery capable of collecting the oppressive warmth of the summer sun and storing it until winter, when you really need it to heat your house. So efficient is this battery that with a properly sized and installed system you will never have to burn another gallon of propane, natural gas or fuel oil to keep your house warm, and—as an added bonus—the Earth also stores the waste heat from your home's air conditioning until the mercury plummets and the snow flies.

Too good to be true? Not at all. Tens of thousands of people are proving every day that it is both practical and cost effective to heat and cool your house with the energy stored in the ground. It's just a matter of moving heat back and forth using the same clever science that makes refrigerators and air conditioners possible: the common heat exchanger.

The Principle of the Heat Exchanger

How is it possible to heat your home cheaply and conveniently by extracting heat from soil that is only 15 or 20 degrees above freezing? By circulating through buried tubing water that is even colder than the earth itself. Since heat always flows toward where it's colder, the water—mixed with an environmentally-safe antifreeze solution—picks up heat from the ground as it circulates. Thus it will be warmer when it returns to your house than when it left it.

The water is still cold, of course; no warmer than the earth it came from. So it has to give up its heat to something even colder—a refrigerant, such as liquid Freon, a substance so cold that it boils at subzero temperatures. Now we're getting somewhere, though it may not seem like it yet. Inside a heat pump unit, the very cold Freon circulates in a double coil with the earth-warmed water, absorbing its heat and making the water cold again, relatively speaking. But even though the Freon has absorbed most of the heat the water gained from the soil (becoming a low-pressure gas in the process), it's still no warmer than the ground outside. And we need to make it hot; hotter even than the air inside our house. How? By compressing it to a very high pressure. This concentrates the gas and raises its temperature to approximately 165 degrees. Then, by running the hot high-pressure gas through a second heat exchanger (either an air duct coil or a hot water tank), the heat is given up into your house. In the meantime, the gas cools to a liquid, runs through an expansion valve, and returns to a cold, low-pressure state, ready for the next go-round.

In summer the process is reversed: the excess heat inside your home is returned to the soil where, as I mentioned earlier, it will be available in a few months for winter heating. And in the process, it cools your home.

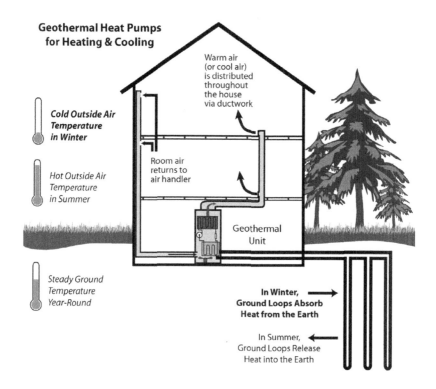

Geothermal Heat Pumps for Heating & Cooling

Warm air (or cool air) is distributed throughout the house via ductwork

Cold Outside Air Temperature in Winter

Hot Outside Air Temperature in Summer

Room air returns to air handler

Geothermal Unit

Steady Ground Temperature Year-Round

In Winter, Ground Loops Absorb Heat from the Earth

In Summer, Ground Loops Release Heat into the Earth

Super-Efficiency

With three separate phases of heat exchange (ground to water; water to Freon; Freon to air or water), the use of a ground-source (geothermal) heat pump to heat and cool your home may not appear to be a very efficient process. But in fact it is extremely efficient—right in the neighborhood of 400%. This means that the ground surrenders four units of heat energy for every one unit of electrical energy you use to extract it. It's like getting three free cords of firewood for every one you buy.

Types of Systems

There are two basic types of underground pipe systems —open loop and closed loop—with several variations. The one you use will depend on where you live, how much land you have, and the characteristics of the soil and ground water.

In a **horizontal closed-loop system**, loops of special heat-conducting polyethylene pipe (either in straight runs or the newer slinky coil method) are buried 6 to 8 feet below the surface. The length of each trench depends on the amount of moisture in the soil. In wet climates, for instance, three trenches, each 100 feet long, are usually sufficient for an average-sized house, while in an arid climate like Colorado's an installer often has to go 200 feet per trench for optimal heat absorption.

Since many of us simply don't have enough land for such a sprawling system, there are vertical closed-loop systems, in which loops of ¾-inch high-density polyethylene pipe are set in concrete in a series of 4½-inch holes. The holes are bored to a depth of 175 to 220 feet, and placed 10 to 15 feet apart. All the separate pipes converge at a manifold, where they are joined into two pipes—one in, one out.

A third incarnation of the closed-loop system is the pond loop. As the name implies, the tubing is floated over a body of water then fitted with ballasts and sunk to the bottom. If you're lucky enough to have a lake nearby, it could spare you the cost of trenching or drilling.

In **open-loop systems**, ground water from a series of wells is used. Water is pumped out of one set of wells, run through a heat exchanger, and

Left: Slinky coils for a horizontal closed-loop GHP system, before burial. Right: a typical geothermal heating/cooling system that also provides hot water. *PHOTOS: GEOSYSTEMS LLC*

is then pumped back into a different set of wells. Since water conducts heat better than dirt, open-loop systems are very efficient and can be less expensive to install than vertical closed loops.

System Size, Cost and Applicability

In a temperate climate, a geothermal system will heat and cool roughly 750 square feet of space per each ton of capacity, which is equal to 12,000 Btu per hour. Installed systems range from $3,000 per ton for horizontal closed-loop systems in ideal soil, to $5,000 per ton for vertical closed-loop systems. If, on the other hand, you have adequate groundwater flow, vertical open-loop systems can save you money.

Are you building a new house? If so, the increase in your mortgage payment from choosing a ground-source heat-pump system over a conventional system will be more than offset by the savings on your utility bill, since a geothermal system will be nearly three times more efficient than any other type of heating and cooling system. If you instead plan to retrofit an existing house, the payback will take 10 to 15 years, depending on numerous factors. The good news is that the equipment should last for 20 to 30 years with little maintenance, while the underground tubing will perform trouble-free for at least 100 years.

Practically speaking, ground-source heat pumps work best with forced-air heating systems, but can also be made to work with hot water heating systems, including radiant-floor heat. When adapted to an existing forced-air setup, the system efficiency can be greatly augmented by sealing gaps and holes in the ducts, adding return-air ducts, and setting up multiple zones. With the addition of a desuper water heater, you can enjoy virtually free hot water in summer and more efficient water heating in winter.

When installed as a hot-water heating system, a geothermal system can only heat water to 110°–120°F, so for most homes it will require some amount of boiler-heated water on really cold days, though overall it should handle 80%–90% of the heating chores over the course of a winter. Summer cooling with ground-source hot-water systems is accomplished by reversing the process. The heat pump unit produces cold water which is circulated to fan coil units which blow air past the cool coils. The cool air is then distributed through a separate duct system.

What is good for the planet is also good for your pocketbook. According to the EPA, geothermal systems are the most energy-efficient, environmentally clean, and cost-effective space conditioning systems available.

There is one caveat: practically speaking, you will have to be connected to the power grid if you intend to install a geothermal heat-pump system. For although the overall efficiency of these systems is unmatched, they still use more electricity than an average house. What they don't use is natural gas, propane, or fuel oil, the traditional fuels that take a bigger and bigger bite out of your energy budget every year.

How much will it cost to heat and cool your home on a yearly basis? I'll give you an actual number: $800. Taking advantage of off-peak electric rates, this is the total annual utility cost incurred by some friends in 2004. And it's not like they live in a small cottage in Texas or a trim little cabin in Georgia. They live with their four children in a 3,500 square-foot custom home in the frigid north woods of Minnesota. All the home's heating and cooling is handled by a closed-loop system using a series of 13 vertical bore holes, while a desuper water heater provides nearly free hot water in summer and very inexpensive hot water the rest of the year. In 2005, when they decided to heat their 860-square-foot garage in winter and turn the house thermostats lower in summer (to appease their teenage daughters), their annual electric bill only went up another $190, in spite of a 3% increase in electric rates. This is about $82 per month for all of a rather large home's considerable energy needs. They calculated system payback time at only three years, and this was before tax credits were available to pay for 30% of the system.

GHP Costs

For what you get for the money, nothing can compare with a geothermal heating and cooling system. So it should come as no surprise to learn that they're not cheap. Nor is there such a thing as a "standard" system; there are several ways to go about extracting heat from the Earth and some are pricier than others.

In his recent book, *The Smart Guide to Geothermal*, author Don Lloyd considers a 4-ton (48,000 Btu), 2-zone system, then breaks down the cost by segments: *Hardware*: $9,000–$13,000; *Labor*: $3,500–$5,000; and *Ductwork*: $3,000–$6,000. The final system cost is determined by which type of piping system is used to gather heat from the ground. In ascending order they are:

- **Open-loop well-source piping**: $2,000–$3,000, not including the cost of the wells.
- **Closed-loop pond- or lake-source piping**: $2,000–$4,000, depending on how far away the pond or lake happens to be.
- **Closed-loop horizontal trenching**: $4,000–$8,000. The size and depth of the trench being the determining factors.
- **Closed-loop vertical installation**: $15,000–$25,000 for drilling and piping.

Confused? Don't be. Most average-sized geothermal systems cost from $20,000 to $30,000. Unless there are peculiar circumstances associated with your system, it will probably fall within this range. And don't forget that there is a 30% federal tax credit if you install one of these before the end of 2016.

....

Air Source Heat Pump

What is an air-source heat pump? Basically, it's a device that uses a refrigerant to extract heat from a cool place and release it in a warm place. A refrigerator is an air-source heat pump that extracts heat from the inside of your ice box (a cool place) and releases it to your kitchen (a warm place). By the same token, an air conditioner transfers heat from your (comparatively cool) house to the great and sizzling outdoors.

Like an air conditioner in reverse, air-source heat pumps are also used to heat buildings by extracting heat from the outside air. Unfortunately, they cannot heat a building nearly as efficiently as ground-source heat pumps, partly because air cannot hold as much heat as water or earth, and partly because the more it's needed—when it's really cold out—the less efficient an air-source heat pump becomes. A ground-source heat pump, by contrast, always has a steady supply of comparatively concentrated heat energy, and does not have to endure the fickleness of an ever-changing atmosphere.

....

Solar Hot Water vs. Geothermal

Is it better to heat your home using solar hot water or geothermal technology? That depends largely on your circumstances, specifically on whether or not you plan to live off the grid. If you do, then a geothermal heat-pump system is in all likelihood beyond the generating capacity of your home solar and/or wind generating systems. But for those of you tethered to the power grid, my advice is to go with a geothermal heating and cooling system. There are two main reasons for this.

In the first place, it just doesn't make sense to invest heavily in a technology that will only be useful during the time of year when it is least effective. Whether you install flat-plate collectors or evacuated tubes, their performance drops whenever the mercury takes a dive. And when the heavy clouds roll in, solar collectors enter into a somnolent state and hibernate—just when you really need them. A geothermal system, on the other hand, couldn't care less whether it is warm or cold, sunny or overcast. You have to admire that kind of indifference in a system.

There's also summertime to consider. For a geothermal system, summer cooling is child's play: just run it backward. It's a trick that doesn't work with a solar hot-water system. But at least with solar you'll get (almost) free domestic hot water, right? A geothermal system can do that, too, with the addition of a desuper water heater.

Secondly, to be truly effective, solar hot-water heating systems require lots and lots of collectors (unlike a domestic hot-water system, which takes only a few). Where are you going to put them all? The most likely answer is on the roof, but a sizable system can easily take up every inch of your home's south-facing roof, and you may need to beef it up to support all that extra weight. You'll also have to find a place for a hot-water storage tank, which can be several hundred gallons in a large system. Contrast this with a geothermal system: except for the heat exchanger, it's all but invisible.

I have friends with both types of systems. Those heating their homes with solar hot water are proud of their systems, and they should be. But they often lament about how much money they have to shell out for supplemental propane. Those using geothermal heating and cooling, by contrast, are continually amazed by how little money they have to spend heating and cooling their homes; just a few bucks for electricity to run the pumps.

Using Solar Electricity to Supplement

A direct grid-tie solar system is a perfect accompaniment to a geothermal system, especially if you live in an area with Time-Of-Use (TOU) billing, where electric rates are cheaper at night than during the day. In this scenario, you will be able to sell your excess solar power to the utility while the sun is shining and the electrical demand is the greatest, and buy it back at night when demand lowest. With a large enough system you can greatly reduce or eliminate your energy bills.

MICK'S MUSINGS

The Cat works a lot like
a heat pump; he concentrates
mischief with super-efficiency.

To learn more about geothermal heating and cooling, or to find a certified installer near you, visit the GeoExchange web site at *www.geoexchange.org* or read *The Smart Guide to Geothermal* (PixyJack Press).

....

Living with a Geothermal System Powered by the Sun

BY DOUG PRATT: As we were planning our new passive-solar home in 2006, we were already looking seriously at geothermal-based heating and cooling. Although these systems tend to be expensive to install, they cost a third less to operate than any other heating or cooling system. When we found that our local small-town heating/cooling contractor was now a fully-trained dealer/installer for these systems, that sealed the deal.

The idea of a geothermal system was attractive from the get-go for its efficiency and low cost of operation, but the fact that I can actually make the fuel—electricity—for a geothermal heat pump was especially compelling, since I obviously can't make the fuel for a propane-fired heater. So heating and cooling costs will be one less thing to worry about if my retirement ever comes.

Summer cooling, which our inland California 100°F summers demand, requires a forced air system, and much as my tootsies would have enjoyed a radiant floor, we just couldn't justify the quite significant extra cost. Besides, radiant floors are

best suited for climates that have constant heating loads. Tons of concrete floor mass don't turn on and off at the flip of a switch. Our passive-solar home would need some heating help on clear cold nights, but very little on the sunny day that would follow. How do you tell all those Btus in the floor to just stay put for 12 hours?

Our system and installation cost about $20,000; $12,000 of that for the pair of 250-foot holes in the backyard. Due to terrain and property lines, a trench-laid plumbing loop would have required some zigzagging, and would have cost as much or more. Because our home is small (1,450 sq. ft.), highly insulated, and passive solar, we were able to get away with only two boreholes. The boring, which was done with a specialized high-speed rig, and installation of the closed-loop plumbing only took a couple days.

How do we like living with geothermal? The most noticeable trait is how unnoticeable it is. The system runs almost silently. Unless you're in the garage, it's practically impossible to tell if the heater or AC is running. Set the thermostat higher, the house gets warmer; set it lower, the house gets cooler. The only time I'm aware of the heater is if I'm seated under a vent when the heater starts up.

Power use is about 2,000 watts, with a rated delivery of 24,000 Btu. To save you the math, that's a little over 7,000 watts output. Heat pumps don't make the energy, they just move it from one place to another. Hence, you seem to get more energy out

than you put in. Pretty slick. And I won't complain about my $100 electric bill since that's for the entire past year and I live in an all-electric house, right down to the electric barbecue grill.

....

– 17 –

Solar Thermal Collectors
Using the Sun to Heat Your Water

Whence Flows My Domestic Hot Water?

Realistically speaking, you have three choices for your domestic hot water supply, any one of which can be incorporated into a solar heated water system. The most common of these is the "good-old" glass-lined tank with a gas burner on the bottom. They're cheap and use no electricity, which is good, but they're all bound to fail in a few years, which is bad. You also end up heating a lot of water that's going to cool off and have to be reheated before you get around to using it, which makes it wasteful.

An **indirect water heater** is another option. These units are supplied with water from the home heating system (the boiler), by way of a heat exchanger. On the plus side, they last far longer than conventional water heaters and, though the initial cost is greater, they will save you money over the long haul. On the minus side, they have the same problem of leaking heat while in standby mode. They also require an electric pump to push water through the heat exchanger.

Tankless on-demand water heaters are your final and best choice. These units heat water in a compact gas-fired burner as you use it. Formerly suited to nothing grander than a weekend cabin, on-demand heaters have gained a lot of sophistication and well-earned respect in recent years. Although the larger models (5-plus gallons per minute) are expensive, they last practically forever and can pay for themselves in a few short years on the energy savings alone. Before you buy, it's important to know that some models will gladly accept pre-warmed water from a solar collection system, while others won't. Be sure of what you're getting. Our Takagi T-K3 tankless heater was more than willing to accept preheated water from our solar-hot-water array, so long as the unit's temperature was set at a scalding 145°F. At lower settings it would often conclude that pre-heated solar water was plenty hot, making for many a cold shower until a Takagi engineer apprised us of a solution.

Solar Domestic Hot Water

Everyone would like to shave a few dollars off their energy bills, and most of us perceive the installation of a solar hot water system as a viable candidate for achieving that end. But what type of system should you install? There are practically as many variations on the basic systems as there are people out there installing them. When LaVonne and I decided to add solar hot water to our home a few years ago, we quickly discovered that nothing is cut and dried with solar hot water—plumbing solar style, it seems, leaves plenty of latitude for the creative spirit to play. Still, there are just a few basic system types to choose from, and that is the best place to start.

Passive Systems for Warm Climates

A **solar batch heater** can best be compared to a big solar camping shower plumbed into your house. It is the simplest and cheapest way to heat water with the sun. A handy homeowner can install one with a small investment and a lot of spare parts and elbow grease. How does it work? Picture the copper pipe that runs from your well, or municipal water supply, to the cold-water supply side of your water heater. Now imagine that same pipe taking a detour through a glass-covered enclosure on your roof or the south side of your house. To complete this mental exercise, imagine cutting the pipe and installing a tank at the exact place where the pipe encounters direct sunlight beneath the glass. The tank—either a commercially available batch tank or something as simple as the liner from a discarded water heater—is filled with water and exposed to the sun through the glass covering the enclosure. Cold water enters the bottom of the tank, hot water is drawn off the top. No pumps, no heat exchangers, no electronic controllers; it's a truly passive system. The pressure to move the water is provided by your home's own water system, and the water heated in the batch tank is the same water that runs through your shower head and your faucets.

It's an attractive idea that works for many, but not for those living in cold climates. The problem is that the piping and the batch tank have to be protected from freezing, something not easy to do in the midst of a northern winter. Batch heaters also suffer from the problem of losing heat at night, which is not very useful if you like to shower in the morning.

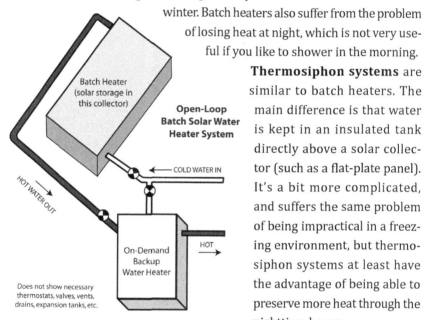

Batch Heater (solar storage in this collector)

Open-Loop Batch Solar Water Heater System

COLD WATER IN

HOT WATER OUT

On-Demand Backup Water Heater

HOT

Does not show necessary thermostats, valves, vents, drains, expansion tanks, etc.

Thermosiphon systems are similar to batch heaters. The main difference is that water is kept in an insulated tank directly above a solar collector (such as a flat-plate panel). It's a bit more complicated, and suffers the same problem of being impractical in a freezing environment, but thermosiphon systems at least have the advantage of being able to preserve more heat through the nighttime hours.

Active Systems for Cold Climates

A popular active system for heating domestic water in cold climates is the **closed-loop drainback system**. As the name implies, the system uses a heat-transfer fluid (usually distilled water) to ferry heat in a closed loop (i.e., a sealed loop that does not come into direct contact with the domestic water supply) from the solar collector(s) through the heat exchanger in a solar storage tank. Potable water moves through a solar storage tank, where it is warmed by the heat exchanger before flowing to the cold-water inlet of the backup water heater.

The pump receives its instructions from a differential thermostat that measures the difference in the heat-transfer fluid temperature between the solar tank and the collector(s). Whenever the tank is hotter than the collector, the pump kicks off and all the fluid drains by gravity flow from the collector into a small drainback reservoir.

These are tried-and-true systems that work well as long as nothing goes haywire with the pump or the controller. Drainback systems must be plumbed with care, of course, since a continuous gravitational gradient is required to ensure that water always drains into the drainback reservoir when the system is inactive, but this can usually be done with just a little forethought. The Achilles heel of this system—from an off-grid point of view—is that it's not pressurized. This means that a much larger pump (and hence, more wattage) is needed to circulate the water between the storage tank and the collectors than would be required in a pressurized system.

Pressurized glycol systems are similar to closed-loop drainback systems in that the heat-transfer fluid (freeze-proof glycol, in this case) circulates through a closed loop, heating water in a solar storage tank via a heat exchanger. The only real difference is that the glycol is pressurized, and thus

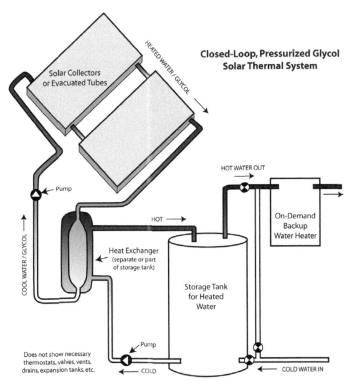

remains in the collectors at all times. This is both good and bad: bad because glycol tends to break down after a few years of service, so it has to be periodically replaced and the system re-pressurized, but good because a smaller pump (usually less than 40 watts) is sufficient to keep the glycol circulating.

....

Anti-Freeze in My Hot Water System?

Isn't anti-freeze the stuff in my car radiator that kills wildlife and the family dog if they drink it? The green stuff in your car's radiator is ethylene glycol, and yes, it's a deadly poison. Worse, it tastes good (take my word for that please). The anti-freeze used by the solar hot water industry is propylene glycol, a close relative of ethylene glycol, but not poisonous. In fact, propylene glycol is FDA-approved as a food additive.

....

Solar Collectors

Flat plate collectors have been around since the 1950s. They are basically flat boxes (usually 4 by 8 feet) covered with tempered glass through which serpentine loops of black copper pipes are set against an absorber plate. They're simple, efficient—and heavy. One Carter-era style collector is all two strong men can handle on flat ground (though newer ones are, thankfully somewhat lighter). *Note: Recycled old flat plate panels without an SRCC rating do not qualify for the federal tax credit.*

Evacuated tubes are the new kids on the block. Each individual unit consists of two glass tubes—one inside the other, separated by a vacuum—that enclose a heat pipe attached to a black copper absorber plate. The heat pipe is filled with a liquid that is converted by sunlight into hot vapor. The vapor rises to the top of the tube where it gives up its heat to a heat conducting manifold through which runs water or glycol.

Which technology is best? It really depends on the application, and how much you're willing to spend. The beauty of evacuated tubes is that they take up less space than flat plate collectors, they're far lighter, and they're less affected by cold or wind. Evacuated tubes perform better in cloudy conditions and when the difference between the (ambient) air and (hot) water temperature is high. For winter heating in cold climates, evacuated tubes can't be beat. On the downside, they are more expensive, so in warm or moderate climates you might do just as well or better with less expensive flat plate collectors.

As with your PV array, the angle of the solar collectors in relation to the sun is important. Since it is highly unlikely that you will make seasonal adjustments of the hot water solar collectors, set the angle for the dead of winter, when you will need most of your hot water. (An

Evacuated solar tube collectors. PHOTO: TOM BORDEAUX

angle of latitude +15 degrees, you will recall, is ideal for capturing winter sunlight. If this isn't feasible, they should, at the very least, be within 15 degrees of latitude one way or the other, and, of course, facing as close to south as you can get them.)

The surface area of the solar collectors is important. The bigger they are, the more water they can heat—that much is intuitive. But the solar collectors will also need to be sized for the tank and the heat exchanger. If the tank is too big or the collectors too small (or too few), the water may not heat up sufficiently to do you much good. On the other hand, if the tank is too small in relation to the solar heating capacity of your collectors you will end up wasting valuable solar radiation because the water in the tank will be quickly heated to capacity during the day but will soon lose much of its heat at night.

The best thing you can do before spending a lot of money on an improperly sized system is to seek advice from someone in your area who routinely installs solar hot water heating systems. Even if they charge you a consulting fee, you'll still be money ahead in the long run.

MICK'S MUSINGS

If a couple of rocking chairs can make the Cat jumpy, I'd love to see what a hot solar collector could do.

What, No Pumps?!

 Solahart and other manufacturers make innovative flat-plate solar hot water systems that depend entirely on thermosiphon action. No controllers, no sensors, no pumps, no moving parts to wear out. The hot water in the collector rises into the storage tank mounted above. These systems are available with open- or closed-loop collector plumbing for temperate or freezing climates. Supplement with an on-demand, tankless heater to ensure you have hot water on cloudy days.

How Much Solar Hot Water Do I Need?

The typical U.S. home uses about 20 gallons of hot water per person, per day for the first two people in the home. After the first two, figure about 15 gallons for each additional person per day. Assuming that your nice hot water will need to be heated from 45° to 125°F—a typical rise—you'll invest about 835 Btus into each gallon, or roughly 15,000 Btus per person per day. If there are four people in your home, your daily hot water energy consumption will be around 60,000 Btu.

Solar collectors are rated by how many Btu per day they'll deliver. What could be simpler? But not so fast. Collector ratings are developed under laboratory conditions, and as we've already (hopefully) learned, it's not a laboratory out there in the real world. In most cases your solar collector is not going to deliver as much energy as the FSEC or SRCC ratings would have us believe. If you're in the Northeast, performance may be as much as 25% less. The collector ratings are useful to compare one collector against another in an apples to apples comparison. Your actual "mileage" may vary.

Solar energy, or irradiation, varies seasonally and regionally. In most of North America a solar hot water system capable of delivering 75%–80% of your hot water needs is ideal. Squeezing out that last 20% requires increasing amounts of hardware with diminishing returns. It's not worth chasing. As a rule of thumb, a good-quality glazed flat plate collector should be SRCC rated to deliver about 1,000 Btu per square foot of surface area on a good day. A common 4-foot by 8-foot collector has 32 square feet. You should expect an SRCC rating at class C (36°F) of 30,000–35,000 Btu per day. So a small household of two people should usually do well with a single 4x8 or 4x10 collector. If you've got more bodies or not a lot of sunshine, you'll need more square feet of collector. System sizing is more art than science, as output depends on regional climate, orientation of the collector(s), and how aggressively you want to go after those "free" Btus. If you're willing to settle for a smaller solar percentage, your Btu return per dollar invested will be higher, but you'll spend more on backup water heating energy. This is a good subject to discuss with your local installer who has a better working knowledge of local solar conditions and knows from experience what can be expected from a particular system.

What Will It Cost to Heat Water with the Sun?

Although costs vary considerably, depending on where you live and what type of energy (electricity, natural gas, etc.) you use to heat your domestic water, one statistic remains fairly constant: you can figure that 15%–20% of your home's annual energy costs will go toward heating water. A properly sized solar water-heating system should be able to supply at least 70% of your domestic hot water, thus reducing, for example, a $650 annual water heating bill to less than $200.

Most active solar hot-water systems for heating domestic water come

Eight of this home's 12 flat plate collectors are mounted on the roof. For much of the year, this large system provides both domestic hot water and home heating, plus enough for an outdoor hot tub. A good use for the old-style collectors that were destined to be recycled.

in for under $10,000 and often for a lot less. Where your system falls in this range will depend on the size of your storage tank and the number collectors you install. Flat-plate panels are cheaper than evacuated tubes, and two to four of them will supply 70%–80% of the hot water needs of most families.

Inline batch heaters are considerably cheaper, and if you build your own you can probably have it up and running for a few hundred dollars. Just don't expect it to perform as well as its glitzier counterpart.

Rebates or Tax Credits for Solar Thermal

As mentioned earlier, a 30% federal residential tax credit is available for most solar hot water systems producing domestic hot water. Heating your pools or hot tub does not qualify. In addition, many individual states or cities offer rebates or credits. Please check your local incentives at the Database of State Incentives for Renewables and Efficiency at *www.dsireusa.org*. You're allowed to double-dip these federal credits, so if you install a PV system and a solar hot water system in the same year, there's no loss of available credit.

A simple 30-evacuated tube system with 80-gallon storage tank and heat exchanger.

Can I Heat My House with Solar Heated Water?

While a good passive solar home design is the very best way to use solar energy to heat your house, a solar hot-water heating system can be adapted to use free energy from the sun to actively augment a passive solar home's efficiency. But in most North American climates that's probably not a smart idea. You're trying to extract a resource (solar heat) at the exact time of year when that resource is in shortest supply. So you'll need a LOT of solar collectors and they're going to sit doing nothing for you seven to nine months of the year.

A solar hot-water system capable of supplying a large fraction of your home's heat is going to cost you a considerable chunk of change. Generally such a system will employ a dozen or more flat-plate panels and a hot-water storage tank of several hundred gallons. The plumbing in these systems is necessarily elaborate and complex with multiple built-in safeguards. All this stuff costs money. Expect to pay anywhere upwards of $15,000 for such a system.

Who Does the Installation?

If you're handy with a pipe wrench and a plumber's torch, chances are you could install a basic batch system without much assistance. But once you get into active systems the level of difficulty rises considerably. Every component in the system must be sized properly, and a number of safeguards will need to be built into it. If nothing else, a consulting fee paid to a seasoned installer could save you a lot of trouble in the long run.

Don't Forget the Tempering Valve!

One problem with all solar hot-water systems is that the water temperature can vary greatly. The water flowing into the cold-water inlet of your backup heater can easily be 180 degrees or hotter—enough to cause a world of misery on bare skin. But with an adjustable tempering valve (also called a mixing valve) between the backup heater and the water tap, it's like having a foolproof thermostat; no matter how hot the incoming water, the tempering valve will automatically mix it with just the right amount of cold water to bring it to a (predictable) preset temperature. Don't jump in the shower without one.

WILLIE'S WARPED WITTICISMS
Solar heated water is great for washing the smell off the Dog, but I'd rather watch him shiver under a cold garden hose.

– *18* –

Biomass (Wood) Heating
Teaching an Old Dog New Tricks

Like geothermal energy, biomass is stored sunshine. The difference is that
the energy in biomass is not stored in the ground as heat, but rather as
the energy in the bonds holding together ubiquitous organic molecules like
cellulose, hemicellulose, lignin, starches and sugars, and so on. All these molecules were formed largely through the interaction of sunlight with water and
atmospheric gases, and they are highly energetic. Pound for pound, a good
hardwood like red oak is four times as energy dense as TNT, which should
help explain why it is so proficient at heating your house.

The other difference between geothermal energy and biomass is that the
energy in biomass is easier to get to. You don't have to bury piping or buy a
heat exchanger; you just have to light a match. And if you burn biomass in an
efficient manner, it will be mostly reduced to ash, water vapor and CO_2, the
latter of which is considered "carbon neutral," as it is composed of carbon
that was already on the Earth's surface to begin with, as opposed to some hot,
dark pool of oil buried underground since the Carboniferous Period.

Biomass comes in the form of cordwood, wood pellets, pellets from forest residues, nut shells, cherry pits, shell corn...you get the picture. There
are many ways to efficiently extract energy from biomass. Most of them are
through technologies for home heating. But not all. Want to learn how run
your pickup or tractor on wood chips? Read on.

*Note: Biomass-burning stoves and fireplace inserts may qualify for a resi-
dential energy-efficient property tax credit of up to $300. Biomass stoves that*

use "plant-derived fuel available on a renewable or recurring basis, including agricultural crops and trees, wood and wood waste and residues (including wood pellets), plants (including aquatic plants), grasses, residues, and fibers" are eligible.

Woodstoves

A fireplace is a wonderful, romantic setting for relaxing on cold winter nights, and if you plan to put one in your new home, I think it's great. There's nothing like the sight of jumping flames to excite the imagination and soothe the spirit.

But if it's heat you want, you will be much better off with a centrally located wood-burning stove or a fireplace insert with large glass doors. It might not be as aesthetically pleasing as a traditional fireplace, but it may save an argument or two over the cost of heating your home.

Although it's an ancient technology, burning wood for heat is still sensible and cost effective. Wood, like wind and sunshine, is a non-depleting source of energy, since most firewood is standing-dead or culled from overgrown forests that needed to be thinned. And with the efficient woodstoves now on the market, wood is a much cleaner fuel than ever before.

The greatest thing about using a woodstove in an off-the-grid home setting is the availability of fuel. Unless your home is in a most unusual place, you should have plenty of wood on, or near, your property. Even if you have to buy it, it'll almost certainly be cheaper than natural gas, propane or heating oil.

If someone offers you a great deal on an old, pre-1988 woodstove, you should, however, respectfully decline the offer. Why? Because 1988 was the year that woodstoves entered the modern world. Concerned about the growing problem of air pollution and woodstoves' contribution to it, the EPA sat down with woodstove manufacturers and kindly asked (as only a government agency can) that all new wood-stoves be designed to meet strict emission standards. The result was a pair of new designs that dramatically reduced emissions and greatly increased efficiency in the bargain.

One of the improved designs utilizes a catalytic converter, similar to the one in your car or pickup, that enables the stove to burn compounds within the smoke that would normally go up the flue and into the atmosphere, unburned. The extra combustion means cleaner air, more heat with less wood, and a stove that can hold a fire longer than any of its predecessors.

Another EPA-approved design, called a secondary combustion stove, accomplishes pretty much the same thing as a catalytic stove without the converter, simply by circulating the gases back through the stove to be burned a second time before exiting up the stove pipe.

Why burn it twice? Because wood does not burn with the same homogeneity as propane or natural gas. Identical masses of wood will give off vastly different amounts of heat, depending on how they are burned.

Here's why: wood has a certain amount of moisture trapped within its cellular matrix, and until that moisture is converted to steam, actual combustion of the wood cannot occur. Since it takes energy to boil water, when you burn wood with a high moisture content you waste a great deal of the wood's energy just getting it hot enough to drive off the moisture so the real combustion can take place.

Once most of the moisture is gone and the wood reaches 540°F the first stage of combustion, called primary combustion, commences. About half of the energy available in the wood is liberated at this stage in the form of heat, and a number of volatile gases, such as methane and methanol, are formed in the process.

We really want these gases to burn, since they contain the other half of the wood's available energy. But it's tricky to do, since secondary combustion occurs at and above 1,100 degrees and requires a lot of oxygen which is generally not available near the burning wood, since the wood is still eating up all the oxygen that comes its way to sustain primary ignition. In a secondary combustion stove, however, the secondary gases are ushered into a special chamber where additional air is introduced to burn these energy-rich gases before they are allowed to exit through the stove pipe.

A catalytic stove, by contrast, avoids the need for a secondary combustion chamber by using a catalytic element which reacts with secondary gases to lower the temperature at which they ignite—all the way down to 600°F. Because of this physical slight-of-hand, catalytic stoves burn cooler and therefore with less intensity than secondary combustion stoves.

Which kind of stove should you buy? Specifications will vary from one manufacturer to the next, but as a general rule catalytic stoves are more efficient, cleaner and more expensive than secondary combustion stoves. Our cast-iron, catalytic stove easily holds a fire all night long, even with fast-burning pine in the fire box. The key is to keep the stove pipe free of buildup—we clean ours twice a season, just to be sure—and the doors adjusted so they close tightly.

High-efficiency, wood-burning fireplaces have blowers to circulate the heat and large glass doors for viewing the cozy fire.

If you have a fireplace that produces a nice, cozy flame but very little in the way of useful heat, a new high-efficiency fireplace insert may be just what you need to turn your wood-burning relic into a useful source of heat. Fireplace inserts are now available in both secondary combustion and catalytic models, and both come with blower fans to distribute heat throughout a large area of your house.

Will a woodstove save you money? At $2.90 per gallon, it takes $478 worth of propane to equal the heat value of one cord of pine firewood. Besides, chopping wood is much better exercise than writing out checks to the propane company.

Masonry Stoves: Heat Batteries

Ask anyone what the purpose of a battery is and they'll tell you it's to store electrical energy for future use. That's the idea behind a masonry stove: to store up heat to be given off slowly over time.

Masonry stoves have been around for hundreds of years. They were originally built as a means to conserve firewood in northern Europe, primarily Scandinavia. The idea behind them is simple: to store as much heat energy as possible from a given amount of wood.

Conventional woodstoves begin to give off heat very soon after the fire is lit. They heat up quickly, and if you're cold it's a nice feeling. In short order you'll find yourself closing the draft and shutting down the damper to keep the room from getting too warm. But by shutting off the stove's air supply you are reducing the combustion efficiency of the burning wood. And even with the stove shut down, a large portion of the heat goes up the chimney and is lost to the great outdoors, since there is not much between the fire and the stovepipe to stop it.

Masonry stoves are just the opposite. They don't rely on the heat-conductive properties of metal to quickly transfer heat to the room; instead, they use the thermal mass of brick and stone to soak up heat and give it back slowly. They do this by routing the flue gases through a long series of baffles built into the structure; baffles that keep the gases moving around and giving up heat every step of the way.

Generally, owners of masonry stoves will burn a single hot fire every one to three days. Once the stove's thermal mass heats up it takes relatively little

A Tulikivi stove made of soapstone is one style of masonry stove. Many are custom-built from local stone. PHOTO: TULIKIVI

wood to keep it warm. The heat given off varies little from one hot fire to the next, since the stove's tremendous mass acts as a constant heat radiator.

Masonry stoves should be built as close to the center of the house as possible, and will perform best in homes with lots of thermal mass to soak up the constant heat. These include log homes, rammed-earth homes, adobe homes, and straw-bale homes—all the really cool homes we're secretly looking for an excuse to build. One drawback is the square footage required for this type of heater, as they have a fairly large footprint. A word of caution: masonry stoves are slow to heat up and slow to cool down, so they work best in places where winter days are cold affairs. If all you want is nighttime heating, buy an ordinary woodstove.

Wood-Fired Boilers

For those of you wishing to lower or eliminate your natural gas, propane, or fuel oil bill, a wood-fired boiler may be the answer. These stout units operate in much the same way as regular boilers, in that they use the heat from burning fuel to heat water running through a heat exchanger. This hot water can then be used for domestic purposes or to heat the home. And since they are designed to burn wood with a very high temperature of combustion (up to 1,800°F; a little above the melting point of silver), a wood-fired boiler can be more efficient than a regular woodstove.

To get the most out of one of these large outdoor units, you should combine it with an insulated hot-water storage tank of several hundred gallons. In this way the excess heat produced above and beyond your immediate needs can be stored for future use, much the way heat is stored within the mass of a masonry stove. And with a storage tank, the boiler can be made to operate in tandem with a solar hot-water system.

Since water is over 800 times denser than air, it is a simple yet highly effective medium for delivering heat where it can do the most good. If you are retrofitting a wood boiler to an existing system, it can be plumbed to supply preheated water to a gas-fired boiler or to bypass the boiler altogether.

When adapted to a forced-air heating system, the hot water from the wood boiler runs through a radiator-like plenum above the blower fan and the moving air absorbs the water's heat as it blows past the plenum into the heating ducts. The water thus cooled by the moving air then circulates back to the boiler through highly insulated cross-linked polyethylene (PEX) tubing for a fresh supply of heat.

Most wood boilers come with thermostat controls on the air intake, so the rate of combustion is constantly regulated to keep your house in the comfort zone without burning more wood than is necessary. Although they only need to be stoked once every day or so, it can still be an inconvenience for those of you who spend a lot of time away from home. For you the answer might be a multi-fuel boiler, one that can automatically switch over to propane, natural gas or heating oil when the temperature in the combustion chamber falls to a preset level. Then when you get home you simply light another fire and the unit switches back to wood-heat mode.

Consumer Concerns

Excessive smoke, resulting from the incomplete combustion of fuel, is one of the most common complaint about wood boilers. This can be caused by a combination of factors, ranging from poor boiler design to inappropriate fuel. To reduce harmful emissions, some areas of the country have passed local ordinances regulating the placement and use of wood boilers. The Environmental Protection Agency (EPA) does not regulate wood boilers, per se, although they have set up a Phase-2 rating program in which a large number of manufacturers have volunteered to participate. The EPA's Burn Wise website *(www. epa.gov/burnwise/)* provides a list of rated models and their manufacturers.

Wood boilers have garnered the praises of thousands of homeowners who have found them an economical and effective way to heat their homes. These are mostly folks who use dry, split, seasoned wood in well-designed boilers properly sized for their homes, since wet wood and/or an oversized boiler will burn too cool most of the time to utilize the abundant energy in secondary combustion gases. Moreover, because the thriftiness of a wood boiler begins to diminish rapidly as wood prices escalate, the really satisfied boiler owners are the ones with plenty of wood nearby that can be obtained at the expense of a lot of time, a few sore muscles, and a bit of chainsaw maintenance.

If you would prefer to stick to a single fuel (and who wouldn't?), HS-Tarm *(www.woodboilers.com)* produces a wood-pellet-fired boiler that automatically feeds itself, much like a pellet stove. All you need to do is fill a bin with pellets and let the boiler take care of itself. (In many places in Europe, where biomass heating is rapidly gaining in popularity, you can phone in a delivery of wood pellets right to your bin, in the same way you order propane.)

A word of caution for off-gridders: since a fan is used to force air into the combustion chamber, these units do require some electricity beyond that used to run the circulating pumps for the hot-water heating system. Check with the manufacturer before you buy and then plan accordingly. Additionally, if you are considering a pellet-fired boiler, extra electricity will be needed to run the auger.

Corn and Pellet Stoves

As a former breeder of race horses, I can attest to the potency of corn as a high-energy fuel (if you have any doubts, read the calorie content for a handful of corn chips). The main advantage of corn—aside from the fact that it is such a clean fuel—is its availability. You can buy corn anywhere people keep livestock; just about everywhere, in other words.

Corn stoves differ from woodstoves in that they have very small combustion chambers and do not require the expensive chimney systems needed by woodstoves; the small amount of exhaust gas is much cleaner and cooler than wood smoke and can be safely vented horizontally through a wall.

You should call a few feed mills to ascertain the local cost of shelled corn (14–15% moisture is ideal) before investing in a corn stove. While almost all of the comparisons I found for the cost-effectiveness of corn stoves listed corn at $2.50 per bushel, you may end up paying a lot more (as I write this, corn in Colorado is $13.00 per bushel). As with anything, the more you buy the cheaper it will be. Your best bet is to buy directly from a farmer in one-ton lots.

Pellet stoves work on the same principle as corn stoves. But instead of corn they burn pellets made from lumber mill scraps, agricultural refuse, or even waste paper and cardboard. American Energy Systems makes a multi-fuel stove that burns corn, pellets, and even cherry and olive pits. Most of these things would have been plowed under or left to rot in bygone days, but now they are considered biomass, a broad classification of environmentally friendly plant-derived fuels that are both renewable and carbon neutral.

Note: Corn and pellet stoves use electricity to run the auger and blower motors. How much? To answer that question, we loaned our Watts Up? meter to Golden Grain Corn Stoves. Set to burn one bushel per day (an average setting), the stove used around 159 watts per hour, or 3.82 kWhs over 24 hours. This exceeds the energy usage of three efficient refrigerators, making these stoves highly impractical for off-the-grid homes.

MICK'S MUSINGS
It's nice to curl up with a warm dog on a cold winter's night. You can think of us as kibble-burning stoves.

Making and Using Wood Gas

The purpose of a secondary combustion woodstove or wood boiler is to extract as much heat energy as possible from a given quantity of wood. It does this in two stages. In the first stage the wood's moisture content is driven off in the form of steam and numerous easily combustible carbon compounds such as cellulose and hemicellulos are converted into heat and secondary combustion gases. Although these remaining gases contain at least 50% of the wood's available energy—mostly in the form or hydrogen (H_2), carbon monoxide (CO),and methane (CH_4),—they cannot burn at temperatures lower that 1,100°F. That is the purpose of a secondary combustion chamber: to provide an oxygen-rich environment capable of sustaining the temperatures necessary for extracting the available heat from secondary combustion gases.

But nowhere is it written that secondary combustion gases have to be burned inside a stove or boiler. In fact, it is sometimes practical to burn them elsewhere, like in the internal combustion chamber of a gasoline motor. This is nothing new; people have been doing it for well over 100 years, ever since internal-combustion engines began replacing external-combustion engines (aka steam engines) at the end of 19th century. During World War II when gasoline was in short supply, wood gas (as it is commonly called) was used to drive stationary generators, farm tractors, trucks and even fishing and ferry boats. In Denmark during the war, where gasoline was as rare as silk stockings, 95% of all motorized transportation ran on wood gas.

Before you attempt to reroute the secondary combustion gases from your woodstove into the intake manifold of your vintage Corvette, however, there is one thing you should know: it won't work; at least not very well. The reason is that "real" wood gas, the stuff that cars and buses and ATVs run on, has to be created in a highly oxygen-deprived environment. Otherwise, during the primary combustion stage, an unacceptable amount of water vapor and CO_2 are produced. True, these gases can be "cracked" into H_2 and CO at high enough temperatures, but the reaction draws heat away from the combustion cycle and if introduced in sufficient quantity they will interfere with secondary burning.

This means that you will need an apparatus other than a woodstove to create your wood gas; you will need to build a wood gas generator. In their most basic incarnations these are gratifyingly simple burners. A download-able 1989 FEMA document, "Construction of a Simplified Wood Gas Generator for Fueling Internal Combustion Engines in a Petroleum Emergency," gives detailed instructions for building one of these units from a garbage can, a small barrel, a mixing bowl, and an assortment of plumbing parts. Even Wile E. Coyote could get it right. Unfortunately, the FEMA design permits the creation considerable amounts of tar, a creosote-laden substance that can quickly gum up a motor. That's why it's designed for a "petroleum emergency." A more recent World Bank document (Technical Paper No. 296) provides a consid-erable amount of information regarding the application and performance of various types of wood gas generators. And, of course, the Internet abounds with construction plans, YouTube videos, anecdotal examples, and even the opportunity to buy ready-built generators.

With so much documentation readily available, you should have no trouble finding detailed plans for the construction of a wood gas generator. The Keith gasifier, designed by Alabama cattle rancher Wayne Keith, is reputed to be the best wood-chip gasifier around. Keith has run his pickup trucks entirely on wood chips for years and is always finding new ways to improve his designs. He's even propelled his truck from coast to coast on nothing but wood chips. His website is *www.driveonwood.com*.

Wood gas generators run primarily on woody refuse: branches, twigs and wood chips are most commonly used, although gasifiers have been made to run on charcoal, rice hulls, wood pellets, nut shells and even coal. The moisture content of the fuel should be kept to 15% or less, which is just a little above the range for kiln-dried framing lumber.

The basic design of a wood-gas generator is simple; as with most things, the details provide the hair-pulling minutiae. The most successful design for woods chips and similar fuel stock is the stratified downdraft gasifier. Basically, fuel is fed into the top and allowed to smolder in an oxygen-restricted environment. As combustion occurs and the fuel stock breaks down, it filters downward by gravity through a series of zones (a drying zone, a distillation zone, and a hearth zone), becoming hotter and ever more anoxic along the way. Wood gas—the result of incomplete combustion—is drawn off just above the reduction zone. From there is it cooled and filtered and ready to be more fully combusted in a high-oxygen environment, such as inside the cylinder of a V-6, or perhaps the burner of an outdoor cooking unit.

Uses and Limitations

Because it is not practical to store wood gas, it should be used as it is produced. The rate of use affects the quality of the gas; if it is drawn off too quickly there may be unacceptable amounts of tar present. So to be on the safe side, gas should not be drawn off the gasifier at a rate more than one-third of its maximum rate of production. This means that for successful operation the gasifier should carefully sized to its application.

And what are the applications? Most plans you will find are for outdoor cooking units in which wood-gas is used much in the same way propane is used in outdoor barbeque grills. Because tars can be burned with impunity in outdoor cookers, the gasifier design need not be particularly refined and extensive filtering of the gas is not necessary.

More serious applications, however, require commensurately more planning in design and construction. Wood gas is used successfully to run gas generators, which can power a house or cabin directly, or to charge a bank of batteries in exactly the same way as a wind turbine or a solar array. Or as previously shown, to run a pickup or any of a number of other conveyances.

How far can you go on a hopper full of wood chips? According to Wayne Keith, 16 pounds of wood will propel his pickup as far as one gallon (6-pounds) of gasoline. Though I have no immediate plans, if I were to convert my truck to run on wood gas I would, according to Keith's calculations, need 64 pounds of wood chips for an average trip to town and back. A good wheelbarrow load, in other words. A bit cumbersome, perhaps, but eminently doable.

But if you've been waiting for me to tell you how you can heat your home, run a propane fridge, and bake a pizza using wood gas, I'm afraid you're about to be disappointed. It simply cannot be used indoors. Because purified wood gas is odorless your nose could not detect a leak. And carbon monoxide is not really something you can live with.

– 19 –

Gotta Have Water

Finding & Storing the Essential Ingredient of Life

This chapter is included for the many readers who have undertaken the task of learning about renewable energy for the purpose of
using it to power a remote home. For those of
you in a more urban setting, the information
presented here may be nothing more than a
curiosity. Just the same, there's lots of good stuff
here on which to exercise your imagination.

Water is your property's most valuable
asset. Electricity you can make; food, firewood,
propane and all other essentials can be brought
in from the outside and stored for long periods
of time. But if your land doesn't have sufficient
potable water, your life will largely revolve
around the transportation and storage of the
liquid of life.

In some places ground water is abundant; in others it's hit-or-miss affair.
Our neighbor to the west got a fair well (3 gallons per minute) at 340 feet,
while our neighbor to the east got a trickle (5 gallons *per hour*) at 700 feet. We
had no idea what to expect, but after watching the driller sink a hole 480 feet
through impermeable rock with no water in sight, any optimism we earlier

felt quickly dissolved. Then, like magic, the morphology of the rock changed and water appeared. Lots of it. By the time the drill bit reached 540 feet, we had a well producing 5 gallons per minute. We let out the breath we'd been holding for several days and uncorked a fine Merlot.

Our problems were far from over, of course. Being off the grid and therefore on a strict energy diet, we still had to figure out the best way to get the water from the bottom of the well to the house. But at least we were dealing with definable parameters. After what we'd just been through, it seemed like a manageable concern.

Just the same, we carefully weighed the options—numerous and varied as they were—before deciding what we'd do.

MICK'S MUSINGS

I've observed that cats don't care much for water. It must dilute the spit and vinegar flowing in their veins.

Should You Install a Cistern?

Before buying a pump, you will need to decide if you are going to pump water into a cistern and then pressurize the house water with a much smaller pump, or simply forget the cistern idea and pressurize the house directly from the well pump. There are three primary reasons people use cisterns: they have low producing wells, their renewable systems aren't powerful enough to run the well pump on a consistent basis, or they want plenty of water on hand for fire protection.

Low Producing Wells

People who have wells with very low recharge rates use cisterns to provide a buffer between what the well can store within its casing and the amount of water they might need to use within a short period of time. As an example, let's say that you have a well with a paltry 5 gallons per hour recharge rate. In one day, it will provide 120 gallons of water. Not much, but enough for two people who are aware of the limitations. But if the well casing only holds 70 or 80 gallons between the static water level and the pump, that's all that can be used in a short period of time. However, by pumping the entire contents of the well casing into a 1,000 or 1,500-gallon cistern every time the well is fully recharged you will be assured of always having enough water even though the well is a poor one.

These systems are common around where we live. A really deluxe setup uses a float system within the cistern (similar to the one in a standard toilet) to automatically turn the well pump on and off. For off-gridders, a switch can even be installed to start a generator to run the pump. A sensor within the well will shut down the pump (and generator) when the water level falls too low.

Not Enough Energy To Run The Well Pump

Many people are reluctant to run an AC deep-well pump with a PV/wind system. This is understandable. Unless you have a robust system with plenty of battery power it's hard to commit that much precious wattage to running a high amperage pump. Others fear they might push the limitations of their inverters. Whether their fears are well-founded or not, most people on PV/wind systems with deep well pumps choose to pump their water into a cistern with a fossil-fuel-fired generator or a stand-alone direct solar-powered DC submersible pump. They then pressurize their house water line with a small AC or DC pump.

We know from experience that it does take a fair bit of energy to run a deep-well submersible pump from the batteries and that it does tax the inverter at times when other heavy loads are running, though never enough to threaten the integrity of the system. We are, after all, pumping water from 540 feet down with an 11-amp, 240-volt pump. Still, we had to try it even though many people told us it wouldn't work. The reason we did it is simple enough: by automatically pressurizing the house water with a deep well pump, it's one less thing to think about, meaning that we never have to worry that we'll be soaped-up in the shower some morning only to discover that one of us (namely me) forgot to fill the cistern. Besides, we don't even like to listen to our neighbor's generator run every night from across the canyon; we like to listen to our own even less.

But don't despair; as you will soon discover, a good DC pumping system can bypass all of the above concerns, though it will open yet another can of worms, as water is a business abounding in worms.

Water For Fire Protection

After years of drought, fire is on everyone's mind around here. Having come dangerously close to losing our home to fire in 2000, 2011—when we drove through wind-driven flames to escape—and 2012, we think about fire a lot. The local volunteer fire department recommends that everyone have at least 2,500 gallons of water stored for fire protection. A large cistern can accommodate a good part of that amount. However, it is doubtful the local fire department will ever tap into your cistern. Our local fire chief recently told me that in over 20 years of fighting fires he has yet to draw a drop from a homeowner's cistern. Heeding that tidbit of information, we instead keep out fire-protection water above ground in a 1,500-gallon agricultural tank outside the house and a pair of ponds that each hold an additional 600 gallons. Should they even need it, the firefighters will have a much easier time finding a pair of ponds and a big green tank than they will an indoor or below-ground cistern that will more than likely not be full when a fire comes roaring through.

Well Pumps

There are two broad categories of submersible well pumps: AC and DC. Both have their strong and weak points, and neither can be used successfully in every application. In most instances, I agree with Demetri, the venerated pump installer who services most of the wells in these hills, when he says that you should never ask a DC pump to do what an AC pump can do better. But I'm getting ahead of myself.

DC Well Pumps

DC-operated well pumps can withstand a range of voltage that would quickly destroy an AC pump. Because of this, they are used primarily in stand-alone systems. This means that you can wire them to their own solar array or wind turbine and forget about them. They will pump water whenever the sun shines or the wind blows. Within certain limits they are designed to work with whatever power is available to them, so long as it's enough to start the pump turning. Many models can be run dry without sustaining damage, which makes them quite useful in wells with low recharge rates. With the addition of a pump controller, the pump can be wired into a float switch that shuts off the power when the cistern is full and starts it again when the water drops below a preset level. Such systems are also used for pumping water for livestock in remote locations.

Admittedly, it's an attractive idea. With a big enough cistern you'll never have to worry about having enough water, even during a cabin-fever-inducing run of non-productive weather. And by having the pump hooked to its own power supply, you won't have to draw from your household energy savings—or drag out the generator—to shower or wash the dishes.

Components of a Stand-Alone Water Pumping System

Most stand-alone solar water pumping systems have just a few basic components. They include a solar array (and/or a wind turbine, both separate from the house system), a pump controller, a well pump, a storage tank, a float valve, a pressure tank, and a booster pump.

The solar array will be sized for your application and can be anywhere from 80 watts to several kilowatts of capacity. The pump controller will be specific for each type of pump. The purpose of the controller is to monitor the power coming in from the array or wind turbine and to adjust the voltage and amperage for optimal pump performance. The controller protects the pump and ensures that there is enough power to run it safely. When power begins to wane, the controller will shut the pump down.

A storage tank is used to store the water needed at night and during cloudy spells. You will want it large enough to store several days' worth of water. It should be fitted with a float switch that tells the pump to shut down when the tank is full. If pressurized water is needed for domestic use, a small inexpensive booster pump powered by the main house PV (photovoltaic) system can be plumbed in tandem with a separate pressure tank. Since water cannot easily be compressed, the air inside the pressure tank acts as a cushion for the booster pump between the kick-on and kick-off pressure settings. This leaves us with the heart of the system, the solar-powered well pump.

Types of Solar-Powered Well Pumps

There are dozens of makes and models of solar-powered well pumps, but all are either surface pumps, which are mounted above the water line and draw water from not much more than 20 feet (the exception being jet pumps, which pump water from deeper down by using extra power to force water through a venturi-like ejector set inside the well); or submersible pumps, which are lowered into the well and can pump water from 800 feet or more.

These can be further divided into two basic types: positive displacement pumps that trap water in cavities and force it upwards, and centrifugal pumps that force water outward with an impeller turning at a very high rpm. In general, positive displacement pumps are used where high pressure is needed directly from the well, and/or in situations not requiring very high volumes of water. In addition, positive displacement pumps can run in low-light conditions, owing to fact that they are able to pump water effectively at low rpm. Centrifugal pumps, on the other hand, are noted for their ability to pump large volumes of water quickly. But they begin to slip (and thus to waste energy) in high-pressure applications and they often cannot develop the power they need to operate in sub-optimal sunlight.

The pump you choose will depend entirely on your application. The idea is to buy the thriftiest pump, in terms of energy usage, that will adequately fill your needs. If, for instance, you are in a situation where you can pump water slowly throughout the day and store it for later use, you might want to look at a Lorentz PS Series helical rotor positive displacement submersible pump. These pumps operate on 24 to 48 volts and can pump several gallons per minute using energy from a solar array as small as 80 watts.

Where more volume is required, as for livestock operations, check out the Grundfos line of SQFlex submersible pumps. They have both helical rotor and centrifugal models, capable of pumping up to 5 gallons per minute (gpm) from 600 feet, or as much as 85 gpm from shallow wells. These highly versatile pumps run on 30 to 300 volts of DC power, or 90 to 240 volts of AC, and can operate on as little as 130 watts of solar input.

If you're lucky enough to have a spring or well under 20 feet in depth, a surface pump might be the better solution to provide your home or garden with water. The Dankoff Solar SlowPump, for instance, is a rotary-vane (positive-displacement) surface pump that can push water far uphill through considerable lengths of piping. These pumps are designed to run optimally on 12 or 24 volts.

Before you lay down good money for a pump or begin gathering any other components for your install-it-and-forget-about-it solar water-pumping system, you owe it to yourself to talk to a dealer who sells several types of pumps. Every application comes with its own peculiarities and by asking you questions you never thought to ask yourself, the dealer will be able to design a solar or wind pumping system tailored for your needs. The best part is, the system design is free.

So the question is, how much is it worth to you to never have to worry about water? A good DC pump is an expensive proposition. And the wire doesn't come cheap either. For example, a 48-volt DC pump, rated at a mere 20% the capacity of our 240-volt AC pump, costs twice as much and requires wire six gauges heavier. And there is also the wattage to consider. All things being equal, it actually takes more total wattage to run a DC deep-well pump than our powerful AC pump; it's just that one uses it up slower than the other.

Of course, at the time we were building (from 1999 to 2001), 540 feet was about the practical limit for DC pumps. And, we were told, at our well depth we should expect certain pump components to fail earlier than normal. The thought of pulling up 540 feet of drop pipe at the whim of a fractious pump motor was enough to drive us firmly into the AC camp.

But with the exponential growth of solar and wind energy, pump technology has made great strides to accommodate a growing market. Today's pumps are far more reliable than they were 11 years ago, and can pump from much greater depths. Our 540-foot well is now well within the range of a good submersible DC pump.

AC Well Pumps

In the race to supply your water, DC pumps are the tortoises: they plod along slowly and ceaselessly. That means, of course, that AC pumps are the hares—on steroids. They're lean and mean and tough. They consume a lot,  but they produce even more. They can pump from virtually any depth the well driller can find water, and you can expect them to last for many, many years without service or replacement.

Since AC pumps move such a high volume of water (our 1.5 hp pump delivers 6 gpm) the total amount of energy they use is surprisingly small. As mentioned above, we figure 200 extra watts of solar capacity is enough to power our well pump. The problem is that an AC pump needs so much power all at one time. Specifically, we're talking about the amperage the pump demands in the split second when it goes from "off" to "on" (the surge), which may be as much as three times higher than its rated amps.

For us it's seldom a problem. But it may be for others. Our well pump operates nicely within the rated capacities of our Trace SW4024 inverter as long as we're mindful of other loads that might be operating at the same time. To be running the clothes washer, a dishwasher, a gas-powered dryer, a stereo system and a couple of computers the instant the pump kicks in has never presented any difficulty. But whenever I (absentmindedly) start up my table saw while everything else is running, the inverter throws in the towel and repays my negligence by providing us with our own personal blackout. Which admittedly is better than the ones the power company used to create back in the last century when we lived on the grid; at least I know how to fix the ones I cause myself.

A larger inverter or two stacked inverters would alleviate the problem, and if you go with a watt-gobbling AC well pump, your PV/wind equipment supplier will almost certainly try to convince you to buy two inverters. The choice is yours. If it turns out you have difficulty running a large well pump with a single inverter, the problem may lie more with the pump than the inverter. This is because all AC well pumps are not created equal; some require more power to start than others (see inset). Your pump installer should be made aware of the limitations of your inverter(s) and should be able to sell you a pump that falls within the parameters set by the inverter manufacturer.

Specifically, you will want a pump that requires a separate starting box outside the well (inside the mechanical room) rather than a pump that has the starting circuitry built into the motor casing. Also, a simple relay-type starting box will work better with an inverter than an electronic one.

Also, if you use a 240-volt transformer to supply power to the well pump it is less work for the inverter if you place the transformer between the pressure switch and the starting box, rather than between the inverter and the pressure switch.

Explain these things to your pump installer. He or she should know exactly what your concerns are and how to remedy any potential problems. If not, there are always other installers down the road.

Grundfos Well Pumps

For a well pump that doesn't "slam" the inverter on startup, check out the Grundfos SQ series high-frequency pumps. They have a soft start, and can be programmed to pump at various rates. For ultimate flexibility, the SQ Flex pumps can run on either AC or DC, across a wide range of voltages.

Rainwater Collection

More and more people are collecting rainwater these days. We use our roof rainfall to keep our fire-protection ponds topped off (and the deer watered), and to fill barrels that water LaVonne's irises and gladiolus (which the deer snack on after drinking from the pond). Just how much water will you collect from your roof? I've come up with a painless formula for determining how much water falls over a given area:

area of roof (sq. ft.) x inches of rain x 0.623 = gallons

(0.623 is derived from 0.0833, the number of feet in an inch, times 7.48, the number of gallons in a cubic foot)

If 0.5 inches of rain falls on a 1,200 square-foot roof, how many gallons are collected? **1200 x 0.5 x 0.623 = 373.8 gallons.** That's a lot of water! You may need more barrels than you thought.

Water by Gravity (for home, livestock, gardens)

Useful as they are, all water pumps suffer from the same set of problems: they cost money to buy, they need energy to run, and they will all someday quit working at an inopportune time. So anything you can do to lessen your dependence on water pumps will work in your favor. Unfortunately, this is not always a practical thing to do, even with proper planning. As discussed earlier, almost everyone with a cistern uses a deep-well pump to fill it. They then use a smaller booster pump to pressurize the house lines. The well pump is an unavoidable fact of life but the booster pump does not always have to be.

Water in a column (or an ocean) builds pressure at the rate of 0.43 pounds per square inch (psi) for every foot of depth. It adds up; at around 2,400 feet below the surface the water pressure is so great that the hull of modern nuclear submarine is in danger of collapsing. At a more modest 150 feet, however, the pressure at the bottom of column of water is over 60 psi. After allowing for pressure lost to friction and constrictions in the lines, you should still end up with around 40 psi, a comfortable water pressure for a sink faucet or a showerhead. Municipalities everywhere use this principle to design and build the water towers you see dotting the landscape in and near cities and towns. By pumping water into a large tank at a suitable height, a community's water pressure can be maintained at a constant rate without the use of additional pumps.

Of course, erecting a 150-foot tower next to your house is highly imprac-tical. But if you live in a place with hilly terrain it might work to your advan-tage to pump water into a large storage tank higher up on the hillside and let gravity pressurize your lines. We have met a number of off-gridders who have done this successfully, either to pressurize their house lines or simply to water their gardens. One Alabama man I interviewed uses pure gravity-fed water from a large limestone cave above his house as his only source of water. On a somewhat less exotic note, our off-grid friends near Wyoming use gravity to pressurize the water lines to their livestock barn. Situated 100 feet below their home's cistern, Phillip and Heather's gravity-induced water pressure is around 30 psi, which is plenty for on-demand livestock waterers and most other uses. If more pressure is needed—perhaps to wash a horse or two—Phillip only has turn a couple of ball valves to send house-pressurized water downhill to the barn.

Gravity-pressurized water systems obviously aren't for everyone. All the people we know who uses gravity to pressurize their systems either have wells located above the house or draw water from a naturally flowing source, such as a spring up on a hillside. It does take more energy to pump water to a greater height, so the higher the source the better. On the plus side, this energy should be offset at least in part by the energy not used to run a booster pump, and the money not spent to buy and install one. That being said, you will still need underground freeze-proof lines to and from the tank and the tank must be insulated in freezing climates, so there will be extra costs incurred.

Another possible (and at this point, theoretical) scheme for getting water uphill is to use excess energy from a wind turbine or a solar array to pump water to an uphill storage tank, perhaps by shunting the unneeded energy through a DC pump controller and then to a DC pump. To be honest, we do not know of anyone who has ever tried it. Should you be the first, I'd love to hear about it.

Unfortunately for us, the highest point on our land sits about 15 feet above our house. Still, it's enough to provide 5 or so pounds of pressure from the 325-gallon tank we placed there to water a few plants and the grass west of the house when Nature is uncooperative.

WILLIE'S WARPED WITTICISMS

If humans were as conservative with water as cats are, all the well-drillers would go out of business.

Decisions, Decisions

It's not easy to weigh all the pros and cons of all the different ways people have conceived to deliver water from the earth under your feet to the sink in your kitchen. Just when you think you have it all figured out, some intractable fact lurking in the shadows jumps out and trips you. Happens to me all the time.

Odd as it may seem, the best remedy for too many facts is more facts. Talk to people on both sides of the issue, AC and DC. Talk to your neighbors. Talk to your well driller and your pump installer. Call the inverter manufacturer. Tell them all what your particular circumstances are. Eventually, the right path will present itself and you'll wonder why you didn't see it all along.

– 20 –

Going On Vacation
Can Your Off-Grid Home Survive Without You?

In December of 2010, as LaVonne and I clung white-knuckled to the sides of a rubber boat swiftly skipping over foamy swells in the Bay of Banderas where we hoped to catch sight of a pod or two of humpback whales, the weather back home in Colorado was taking a sharp turn for the worse. A massive system dropping down out of the arctic was bringing with it subzero temperatures and three straight days of snow.

Had we been hooked to the electrical grid, we doubtless would have patted each other on the back for a rare feat of good timing, since we obviously would rather be basking bare-skinned under the warm Mexican sun than shoveling snow and shivering in our mukluks. However, since the last power pole on our road remains a considerable distance from our house, and said house was 1,360 miles due north of where we currently were, the imminent cold front presented something of a problem.

In winter our house is heated with wood, at least when someone is around to stoke the stove. But we obviously weren't home to tend to that most vital of chores so the house was instead warmed by the backup heating system, the in-floor hot-water setup powered by a propane-fired boiler and a series of zone pumps, all of which require a steady supply of electricity.

Normally, the wattage provided by the solar array and wind turbine would be more than enough electricity to run the heating system as well as the electric fridge and freezer, even during a stretch of cold, cloudy weather. But the weather was beyond cold, the wind wasn't blowing, and the solar array was all but put out of commission by the steadily falling snow obscuring any incoming UV radiation. It all added up to an unusually high electrical load that had to be met solely by a couple dozen L-16 solar batteries.

Fortunately, we'd left our house and dogs in the able care of a nearby neighbor who keeps a rusty old Jeep Wagoneer chained up and fitted with a plow for those rare times when Nature gets a little frisky. Thus equipped, he dutifully lumbered up the steep one-mile rutted road from his home by the creek each morning and evening to sweep the snow off our solar array and check on the status of the batteries. LaVonne and I, being trapped in a tropical paradise, could do little more than read his email weather and battery-status updates and walk him through the intricacies of a charging system comprising two separate battery banks and enough manual switches to confound a fighter pilot.

Happily, it all worked out. Both battery banks were still above 50% capacity by the time the weather finally broke, and they quickly charged back up again once the sun and wind returned. Our neighbor didn't even have to run the backup generator which, in any case, would have been a whole other can of worms for someone unfamiliar with our system.

Things could have been a lot worse. If there is something to be learned from this story, it is that our troubles were the result of our solar, wind and home heating systems being somewhere in the no-man's land between low-tech and high-tech. Had our backup heating system been more primitive, the batteries could have ridden out the storm with nary a care. On the other hand, a sophisticated backup battery-charging system would certainly have made our vacation more of a vacation.

As a matter of both principle and personal economics, most of our off-grid friends favor the low-tech approach to heating their homes when no one's around to stuff wood in the stove: a simple propane space heater. Wall-mounted heaters are relatively cheap and they're really quite efficient at turning propane into usable heat. So efficient, in fact that many models do not even need to be vented to the outside. Best of all for those of us whose homes can sometimes

WILLIE'S WARPED WITTICISMS

Cats are perfectly self-reliant when humans go on vacation. Dogs are more helpless than children.

be energy challenged when we're away, these heaters require no electricity. The downside is that, without fans or ductwork to push the heat from room to room, the warmth they produce doesn't stray very far from the source. But if you are willing to part with a few more watt hours in your absence, you can do as others have done and install a DC ceiling fan to help distribute the warm air. These fans draw only four watts on the lowest setting.

Fan or no fan, one or two strategically located propane wall-mounted heaters can keep the pipes from freezing in your absence, or maintain a steady indoor temperature on chilly nights. We have a 10,000 Btu/hour ventless Pro-Com blue-flame heater mounted in our 550 square-foot guest cabin to keep the stored water from freezing. On the lowest setting it keeps the place near 50 degrees on the coldest nights.

There is, however, another course of action that saves energy and practically guarantees peace of mind; you can let the house freeze after winterizing it. By draining the pipes and water heater and pouring propylene glycol (RV-type antifreeze) into the sink, shower and toilet traps, you can simply let Nature have her way. I used to do this routinely when I lived in a small cabin that had grid power but lacked propane service to run a backup heating system to supplement the central woodstove. A few of our friends still resort to this Spartan approach to energy conservation. Not only does it save money, it promises that the act of returning home in the dead of winter will be a memorable character-building experience.

The second part of the low-tech strategy is to keep the compulsory house loads to a bare minimum. Turn off every non-essential thing that could possibly draw a load in your absence, including cable boxes, clocks and power strips. You should also trip all the ground-fault interrupts (GFIs) in your home; they each draw around five continuous watts.

After the home's heating system, electric refrigerators and freezers are the biggest energy users when no one is home to run the computer, washing machine or shop tools. New energy-efficient fridges and freezers each require a little more than 0.5 kWh per day when sitting idle in a cold house, while older units can demand considerably more wattage. If you don't want to sink a lot of money into a battery bank capable of supplying that much power through a long stretch of frigid, dreary weather, you can always go with a propane-fired model. There are several propane refrigerators out there, but in our experience Servel has always been the brand of choice for off-gridders.

The low-tech approach is not for everyone, of course. I'm thinking now of another friend across the canyon. His 1,800 square-foot off-grid log home is powered by a 3,840-watt solar array and a bank of 24 monstrous 2-volt batteries. In his case, a couple of propane wall heaters just aren't going to cut it. With a full basement he's heating 3,600 square feet, so like it or not, he's pretty much married to his in-floor hot-water heating system. And his rather large electric fridge. Yet when I talked to him recently, he made it quite clear he has no special concerns about leaving home in winter. That's because his Xantrex

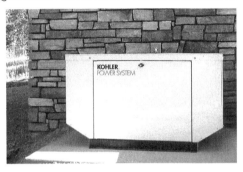

4048 inverter is hardwired to a 10,000-watt Kohler propane-fired generator *(photo)*, permanently mounted on a concrete pad beside his house. Whenever the batteries get drawn down to a preset low-voltage threshold, the generator kicks in and charges them until they reach the absorption voltage. It then holds the batteries at that voltage for a predetermined time (10 minutes, in our friend's case) before the inverter shuts off the generator. It's just like being hooked to the power grid only better, since there are no overhead power lines for Nature to get playful with.

However you go about it—low-tech or high-tech—it's good to know your off-grid home will be safe in your absence whenever duty calls, or irresistible recreational urges overwhelm you in the dead of winter.

MICK'S MUSINGS

Dogs guard the house
when the humans are away.
Cats shred the furniture.

Afterword

So, here you have it. Most of the knowledge I've gained through hard-won experience is in this book, as well as a great deal of what I've learned simply because my curiosity would permit me no rest if I hadn't. But for all that, this book is still just an introduction to the burgeoning subject of renewable energy. Even if you've read every page, you still have a great deal to learn—we all do. Much of it you'll find in articles and books far more technical than this one, or scattered about in a million nooks and crannies in the nebulous maze of cyberspace. But the real pearls of knowledge—those that bury themselves so deeply in your brain that they can never be dislodged—will be those you discover for yourself.

If, as I hope, I was able to convey in these pages the enthusiasm I feel for the prospects of solar, wind and microhydro power, it is because my enthusiasm is real; real because these non-depleting sources of energy have for me become not only a means to an end, but a way of life. I cannot imagine waking up in the morning without parting the drapes and studying the sky, or watching the treetops to see how hard the wind is blowing, and from what direction. Or feeling a tickling urge to step outside to feel the brisk morning air against my face, so I can get a sense for what Mother Nature is serving up today.

To live in step with the rhythms of nature, I have learned over the years, is more than just sharing in the splendor of every bright, sunny day. It is also to acquiesce to being humbled by nature; to realize you are without redress, and to accept whatever comes your way, even if it seems—as it often does—that you are being treated unfairly. The sun shines where it will. The wind blows as it chooses, but no one knows when, or where, or how strong—or for how long. And still you have to believe that the fickle sky will provide for you.

That's the beauty and the paradox of this whole enterprise.

Acknowledgments

It would seem by the third go 'round that I would have run out of people to thank for helping to make *Power with Nature* the book that it is. But such is not the case. As with the previous two editions it was a group effort, and although the group has grown more exclusive it only means the workload has intensified.

Certainly LaVonne, my editor, designer, and publisher has had to work harder on this third edition than either of the previous two. The addition of so much new material necessitated a complete revamping of the book's organizational structure, and this she did with such finesse that what began as a haphazard patchwork now appears natural and seamless. And as her other skills have only grown sharper over the years, their continued refinement requires from her a tireless pursuit of perfection, subtlety demonstrated on every page of the book. Nor would *Power with Nature* be what it is without her continued encouragement and gentle prodding.

I would also like to thank my good friend and *Got Sun? Go Solar* coauthor, Doug Pratt for poring over the manuscript and making numerous comments on its technical merit. Renewable-energy technologies are advancing at a dizzying rate and Doug is one those rare individuals who not only knows what is happening on the far horizon, but understands what he sees and takes it all in stride. There is a lot of Doug in this book.

For help in past editions, I am still greatly indebted to the late Mark "Dr. PV" McCray; former ASES chair, Ronal Larson; and OutBack Power Systems' pioneers Marty Spence, Robin Gudgel and Christopher Freitas. Thanks, guys; I couldn't have done it without you.

As always, I take full credit for all blunders and oversights.

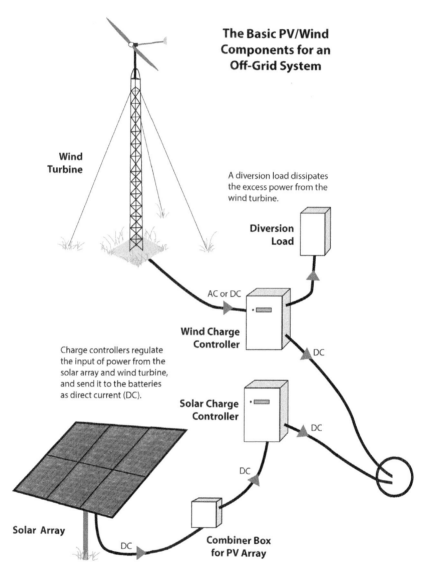

The Basic PV/Wind Components for an Off-Grid System

Wind Turbine

A diversion load dissipates the excess power from the wind turbine.

Diversion Load

AC or DC

Wind Charge Controller

DC

Charge controllers regulate the input of power from the solar array and wind turbine, and send it to the batteries as direct current (DC).

Solar Charge Controller

DC

DC

Solar Array

DC

Combiner Box for PV Array

Solar modules are mounted in an array, and the DC current from all the modules is combined in a combiner box.

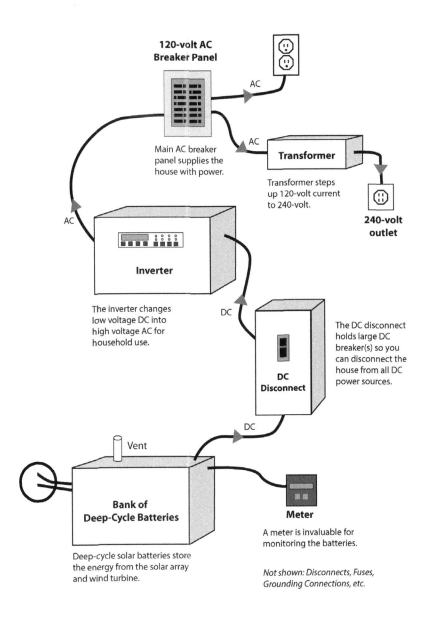

120-volt AC Breaker Panel

Main AC breaker panel supplies the house with power.

AC

Transformer

Transformer steps up 120-volt current to 240-volt.

240-volt outlet

AC

Inverter

The inverter changes low voltage DC into high voltage AC for household use.

DC

DC Disconnect

The DC disconnect holds large DC breaker(s) so you can disconnect the house from all DC power sources.

DC

Vent

Bank of Deep-Cycle Batteries

Deep-cycle solar batteries store the energy from the solar array and wind turbine.

Meter

A meter is invaluable for monitoring the batteries.

Not shown: Disconnects, Fuses, Grounding Connections, etc.

Solar Cells & MPPT Charge Controllers:
What Makes Them Tick

If you've ever wondered what really goes on inside a solar-electric cell or an MPPT charge controller (or direct-tie inverter), below is a fairly painless explanation for each process.

Solar Cells

Largely because of the ubiquity of computers, anyone who has not been living in a cave for the past two decades knows that silicon is a semiconductor of electricity—it sort of allows an electrical current to pass through it, but hardly with the facility of copper or aluminum. Oddly enough, it's this quasi-standoffish attitude of silicon that makes it so useful in the manufacture of solar cells.

Chemically, silicon has 14 positively charged protons, and 14 negatively charged electrons. This would seem to be a happy arrangement, if not for the fact that it has room for four more electrons in its outer energy level. How does it get them? It could snatch four passing electrons from somewhere, but there would be no protons to hold them in place, so the kidnapped electrons would soon escape. So it borrows them from other silicon atoms, forming a crystal lattice in the process. In this crystal, every atom of silicon is attached to four other atoms of silicon and they all share electrons. In other words, every silicon atom has the four extra electrons it wants with no net charge, since the protons in the crystal exactly balance out the electrons. It's a really cushy setup.

Silicon Doping: Homogeneity's Undoing. In fact, it's far too cushy for our purposes. Happy silicon with happy electrons is pretty useless if we want it to do any work. We need to stir things up a bit. How? By adding impurities to it. Say, a few atoms of phosphorus. Phosphorus has five electrons in its outer energy level, so if it is introduced into the silicon crystal lattice (in a process called doping), that fifth electron will be frantically looking for a place to fit in. Now we have an unhappy electron and that in turn makes us happy.

But we're just getting started. A melancholy electron wandering aimlessly in search of a home doesn't do us much good. We need to give this electron a purpose—something it can aspire to. We do this by doping the other side of the silicon crystal, this time with a different impurity, say boron. Having only three electrons in its outer energy level, a boron-doped silicon crystal will have empty spaces where electrons could be, but aren't. These empty spaces are called holes, and each of these holes would like to have an electron to call its own.

Are you beginning to see where this is leading? Our phosphorus-doped silicon is called n-type, in honor of the extra negative electrons, and the boron-doped silicon is called p-type for the extra positive holes. And we're getting very close to having a useful electronic device.

Life at the P-N Junction. At first blush you might think that all the extra electrons in the n-type silicon would zoom across the p-n junction (the place where the two opposite types of silicon meet) and fill in all the holes in the p-type silicon, but it just doesn't happen that way. Oh, a lot of them start out fast enough but quickly begin to have second thoughts. Sure, an electron soon realizes, there may be a nice cozy place for me on the p-side, but my faithful proton is still on the n-side. I'm so confused. It's a bit like young love.

The important thing to remember is that, even though n-type and p-type silicon have extra electrons and holes, neither type alone has any net electric charge. In both cases there are just enough electrons to balance out the protons. But once the rush across the border occurs, that quickly changes. Every time an n-type electron jumps through the p-n junction and fills in a p-type hole, a negative charge is created on the p-side, while a positive charge springs up on the n-side in the place where the electron was, but no longer is. Once everything settles down, we find that there is a great gathering at the p-n junction, with negative electrons lining up along the p-side, and positive holes lining up along the n-side. This creates an electrical equilibrium, and if we left things alone the p-n junction would be a really boring place.

But we're not through yet, for now it's time to finish building our solar cell. To do this, we need to crisscross the surfaces of our silicon wafer with electrically conducting channels. This will provide an easy path for the electrons to travel along, once we add the magical ingredient, sunlight.

When a photon of light of the right energy and wavelength strikes an electron hanging out with all of his buddies on the p-side of the p-n junction, the electron is instantly imbued with a jolt of energy and is suddenly free to move around. Where does it go? It can't go any farther into the p-side; there's quite a crowd there already. So instead it uses all this free energy to make a beeline back to the n-side. And, with a little luck, it will be picked up by one of the conductors on the surface of the n-layer and sent through an electrical circuit.

A Loopy Idea. Once the process begins, the electrical equilibrium at the p-n junction is hopelessly undone and the proverbial floodgates are opened. In an instant, multitudes of electrons that were just moments before hanging out at the p-n junction are whisked out of their silicon Shangri La, drawn through the windings of a washing machine motor or the filament of a light bulb, and unceremoniously dumped back on the p-side of the solar cell, totally exhausted. But, like battered and beaten heroes in a video game, all they really need is a little nourishment—a single photon—to be right back in the thick of things.

The completed circuit is the key to making the whole thing work. Since a solar cell acts as a diode—only letting the current flow from the p-side to the n-side—it wouldn't produce electricity for very long without a fresh supply of electrons continually re-entering the solar cell from the p-side. That's why all solar modules have positive and negative terminals. The electrons flow out of the negative terminal which conducts electrons from the n-type silicon, through the load (the above-mentioned washing machine or light bulb) and back into the p-type silicon via the positive terminal. It's a little drama that's played quintillions of times per second in an average solar cell.

MPPT Charge Controllers

You've been reading a lot about Maximum Power Point Tracking, but what, exactly, is the power point that's being tracked? To find the answer, you'll have to imagine playing a game—a board game. Unlike checkers or Monopoly, however, the board is completely black, except for faint calibrations on the left side and bottom. The left-side marks represent amperage; the marks across the bottom are for voltage.

The board is plugged into your 12-volt array. You have been assigned the role of Charge Controller. You firmly grasp your stylus and flip the switch. A thin bright line appears across the board; a power curve. It's sunny out and nearly midday, so the line runs almost straight from near the top of the board to the right, all the way over to where it meets the array's rated voltage, around 17 volts. From there it curves down and meets the edge of the board at the open circuit mark of 21 volts on the bottom right side.

You place your stylus on a point on the line. Instantly the board lights up below and to the left of your stylus. The lighted area represents wattage, and it's your job to light up as much area as possible. To make it easier for you, a small digital readout appears to show you how much area of the board is lit up. You notice that, as you move your stylus up and down the curve, the area changes. If you go too low on the curve your voltage increases, but your amps are so low that the total power begins to diminish. Move up the curve, past where it straightens out, and you find that you've got lots of amperage, but too little voltage to maximize the wattage.

Finally you find the optimal spot. Pat yourself on the back; you've just tracked the power point. If you do this all day with a constantly changing power curve—as the clouds come and go, as the sun moves across the sky and the array heats up and cools off—you'll know just what it feels like to be an MPPT charge controller.

Of course, if you wanted to be a regular charge controller you'd just touch your stylus on the power curve directly above the battery voltage and call it good. Easy, but not much fun.

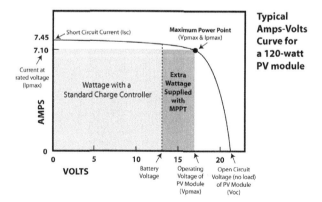

Typical Amps-Volts Curve for a 120-watt PV module

Power Consumption of Appliances

Measured with a Watts Up? Meter

Appliance	Continuous Draw (Watts)
Computer, desktop	80
Computer, laptop	24
22" LCD (flat screen) monitor	50
HP LaserJet printer (in use)	600
HP Inkjet printer (in use)	15
Microwave	1,400
Coffee Maker	900
Toaster, 2-slice	750
Amana Range (propane): Burners	0
Oven (with glow bar; when heating)	380
Blender	350
Mixer	120
Slow Cooker (high/low)	240 / 180
20" Television	50
27" Television	120
27" LCD Television	95
50" LCD Television	190
50" Plasma Television	330
Stereo System	25
Stereo, portable	10
Vacuum, Oreck	410
Vacuum, Dirt Devil Upright	980
Table top fountain	5
Sewing Machine (Bernina)	70
Serger (Pfaff)	140
Clothes Dryer (propane)	300
Clothes Iron	1200
Hair Curling Iron	55
Hair Dryer (high/low)	1,500 / 400
Furnace Fan ($1/_3$ hp / $1/_2$ hp)	700 / 875
Corn Stove	159
Treadmill (walking/running)	150 / 800
Nautilus Elliptical Trainer	4

One watt delivered for
one hour = **one watt-hour**

1,000 watt-hours = one kWh

amps x volts = watts
2 amps x 120 volts = 240 watts

For the latest ratings of
refrigerators, freezers and
other large appliiances, look
at **EnergyStar.gov** website.
Efficiencies keep improving
every year.

Appliance	Watt Hours
Dishwasher, cool dry	736 watt-hours/load
Clothes Washer (front-loading)	145 watt-hours/load

Notes on Appliances for Off-Grid Living

Compact fluorescent light bulbs add up to big energy savings. If 6 bulbs are on for 5 hours a day: 60-watt incandescent bulbs will use 1,800 watt hours per day; 13-watt compact fluorescent bulbs will use only 390 watt hours. CFLs are now readily available in a variety of color temperatures (warm light, cool light, etc.).

Low-usage, high energy appliances (hair dryers, microwaves, coffee makers, etc.) are not much of a problem since they draw very little power when averaged out over time. You can also choose not to use them if you're low on power.

Invest in a new refrigerator and/or freezer. You'll be amazed at how much more energy-efficient they are. The typical new fridge now uses 80% less energy than models from the late 1980s and early 1990s. Research on *www. energystar.gov* before buying and always read those yellow tags!

For a cooking range, buy one without a glow bar, which can use 300-400 watts ALL the time your oven is on. Peerless-Premier is one brand that is ideal for off-grid homes.

To conserve energy and water when washing clothes, a **front-loading clothes washer** is a must, as is a gas-fired clothes dryer. Better yet, use a clothesline or indoor rack for drying.

Instant (on-demand) water heaters, either gas or electric models, use 20 to 40% less energy because they only work when someone turns on the hot water faucet. They also last 30 to 40 years, reducing landfill and resource waste.

Combine a **solar hot water** system with an instant water heater and you've got the lowest-cost, most ecologically responsible way to heat domestic water.

Energy Conversions

- Btu (British Thermal Unit): the energy required to raise one pound of water one degree (F).
- 1 gal. liquid propane = 4 lb. (if you buy propane by the pound)
- 1 gal. liquid propane = 91,500 Btu
- 1 gal. liquid propane = 36.3 cubic feet propane gas @ sea level
- 1 Therm natural gas = 100,000 Btu
- 1 cubic foot natural gas = 1,000 Btu
- 1 kW electricity = 3,414.4 Btu/hr
- 1 horsepower = 2,547 Btu/hr

Source: Solar Living Sourcebook

System Sizing

Shopping for a PV/Wind System

When we wrote *Logs, Wind and Sun,* we thought it would be an interesting exercise to request bids from four companies as to what type of system they would recommend given the variables listed on the following page. Three companies promptly responded to our request; the 4th one took a few weeks. Their responses varied greatly—from asking many questions to making assumptions; from providing very detailed bids to giving an approximate package price.

Their equipment recommendations varied too, from 600 watts to 1,440 watts of PV modules (we had 1,140 when surveyed); 2 inverters (we use one inverter plus a transformer); battery capacity ranged from 1,680-amp hours to 2,340-amp hours (we used the 1,100-amp hour bank for over a year, then added a 2nd one).

Some of the questions they asked: new or existing home, grid-tied or off-the-grid; what elevation; how far from home to solar array; how many gallons of water used per day, wattage of circulating pumps and hours used per day; how many light bulbs and usage per day; number of sunny days per week or month.

Getting the prices for the main equipment is easy, but don't forget to include all the extras: battery cables, fuses and disconnects, wiring, lightning arrestors, shipping, etc. If you are comparing apples to apples, an itemized bid is the only way to go if you want to avoid surprises in the end.

Pick a reputable company (get references from satisfied customers); someone who can answer your questions and will still be in business when you have more questions or need warranty service.

Steps To Sizing Your System

1. Determine your power consumption (use worksheet below).
2. Re-evaluate your consumption to look for ways to conserve. *A rule of thumb that you may hear is that for every dollar you spend replacing inefficient appliances, you'll save $3 in the cost of a renewable energy system.*
3. Find your sun hours per day for your location (see maps/website).
4. Size your solar array (use worksheet below).
5. Size your battery bank (use worksheet below).

Have this information ready before you request bids. The companies will know you've done your homework and that you are serious about renewable energy.

Our Solar/Wind System

- Initially: 1,140-watt PV array, tilted seasonally; now expanded to 2,320 watts
- 1,000-watt wind turbine on 50-foot tower
- Trace SW4024 inverter (24-volt)
- Trace 240-volt transformer (to run the well pump)
- Two charge controllers: Bergey Power Center (wind); OutBack FM80 (solar)
- Two TriMetric meters
- Batteries: 2 banks of twelve L-16s (9,360 amp hours total)
- Trace 250-amp DC Disconnects

Our Home

- Log home is 900 sq. ft. on the main level plus 600 sq. ft. of open loft (no radiant heat); plus a 1st level basement/garage of 900 sq. ft. (with radiant heat; temp set at 40 degrees)
- Woodstove used extensively for heating the main floor/loft
- Propane used for on-demand water heater, radiant heat boiler (if needed), cooking range, clothes dryer

Our Power Consumption

The PV/Wind system powers the following:
- Well pump (1.5 hp; draws 11 amps at 240 volts; pulls water from 540 feet at 6 gal/minute; no cistern; 40-gallon pressure tank)
- Radiant heat circulation pumps (5 zones)
- 19 c.f. Kenmore refrigerator (top freezer); plus 5.5 c.f. chest freezer
- Front-loading clothes washer (3 loads/week)
- Dishwasher (2 or 3 cycles per week)
- Microwave, toaster, coffee maker, etc.
- 27" LCD TV (2 to 3 hr/day)
- Desktop computer, 22" flat-screen monitor & many peripherals, laser and inkjet printers (6 to 8 hr/day)
- Laptop computer (6 hr/day)
- Stereo
- All compact fluorescent light bulbs and lots of natural daylight

Comparison of Popular Wind Turbines

	Bergey Windpower XL.1	SouthWest Windpower Whisper 200	Kestrel e300[i]	Southwest Windpower Skystream3.7
Rated Power	1.0 kW	1.0 kW	1.0 kW	2.1 kW
Cut-in wind speed	5.6 mph	7.0 mph	5.6 mph	6.7 mph
Rated wind speed	24.6 mph	26 mph	21 mph	29 mph
RPM @ rated output	490 rpm	900 rpm	650 rpm	330 rpm
Approx. monthly kWhs @ 12 mph	188 kWh	158 kWh	200 kWh	350 kWh
Rotor Diameter	8.2 feet	9.0 feet	9.8 feet	12.0 feet
Maximum design wind speed	120 mph	120 mph	120 mph	140 mph
Turbine Weight	75 lb.	65 lb.	165 lb.	170 lb.
Direct Grid-Tie	no	yes	yes	yes

	Proven WT2500	Kestrel e400[i]	Bergey Excel
Rated Power	2.5 kW	3.0 kW	7.5 kW (DC) 10 kW (AC)
Cut-in wind speed	5.6 mph	5.6 mph	8.0 mph
Rated wind speed	26 mph	24.6 mph	31 mph
RPM @ rated output	300 rpm	520 rpm	310 rpm
Approx. monthly kWhs@ 12 mph	415 kWh	340 kWh	900 kWh (DC) 1,090 kWh (AC)
Rotor Diameter	11.5 feet	13.1 feet	23.0 feet
Maximum design wind speed	145 mph	120 mph	125 mph
Turbine Weight	418 lb.	506 lb.	1,050 lb.
Direct Grid-Tie	yes	yes	yes

Not all turbine manufacturers or models are shown above.

To view maps of average wind speed in your location, visit:
www.windpoweringamerica.gov/windmaps/
and select Residential-Scale Wind Resource Maps

Metric Conversions

mph x 1.6 = kph   mph x 0.447 = m/sec   ft x .3048 = m   lbs x 2.2 = kg

Power Consumption Worksheet

We recommend doing two calculations: one for winter, and one for summer.

Electrical Device	Wattage (volts x amps)	X	Hours of Daily Use	X	Days Used per Week	÷ 7	=	Ave. Daily Watt-Hours
		X		X		÷ 7	=	
		X		X		÷ 7	=	
		X		X		÷ 7	=	
		X		X		÷ 7	=	
		X		X		÷ 7	=	
		X		X		÷ 7	=	
		X		X		÷ 7	=	
		X		X		÷ 7	=	
		X		X		÷ 7	=	
		X		X		÷ 7	=	
		X		X		÷ 7	=	
		X		X		÷ 7	=	
		X		X		÷ 7	=	
		X		X		÷ 7	=	

Total Average Watt-Hours per Day

15% Loss Correction Factor*   x 1.15

Adjusted Average Watt-Hours per Day

*15% to 25% is added to compensate for inefficiencies in the system (batteries, inverter, line loss).

Solar Irradiance Maps for Winter & Summer

http://rredc.nrel.gov/solar/old_data/nsrdb/redbook/atlas

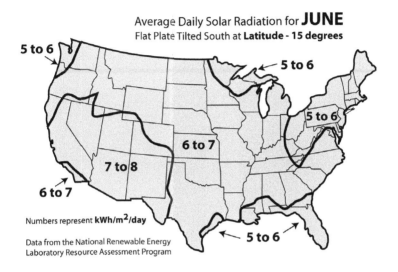

Average Daily Solar Radiation for **JUNE**
Flat Plate Tilted South at **Latitude - 15 degrees**

5 to 6
5 to 6
5 to 6
6 to 7
7 to 8
6 to 7
5 to 6

Numbers represent **kWh/m²/day**

Data from the National Renewable Energy
Laboratory Resource Assessment Program

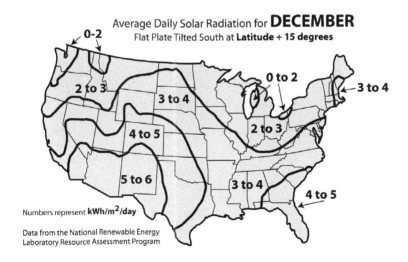

Average Daily Solar Radiation for **DECEMBER**
Flat Plate Tilted South at **Latitude + 15 degrees**

0-2
0 to 2
3 to 4
2 to 3
3 to 4
2 to 3
4 to 5
5 to 6
3 to 4
4 to 5

Numbers represent **kWh/m²/day**

Data from the National Renewable Energy
Laboratory Resource Assessment Program

Solar Array Sizing Worksheet

	June	Example	December
1. Input your Adjusted Ave. Watt-Hours per Day *(from the Power Consumption Worksheet)*	_____	*3000*	_____
2. Find your site on the June and December Solar Radiation Maps (or website) and input the nearest figure.	_____	*6*	_____
3. To find the number of watts you need to generate per hour of full sun, divide line 1 by line 2.	_____	*500*	_____
4. Select a solar module and multiply its rated wattage by .70 (.80 if using an MPPT charge controller).* *Example: Enter 84 for a 120-watt module (or 96 with MPPT).**	_____	*84*	_____
5. To find the number of modules needed, divide line 3 by line 4. *Remember that modules will be wired in series of 2 or more, depending on system voltage.*	_____	*6*	_____

This worksheet assumes you will operate entirely on solar power. . We suggest starting conservatively, and add more solar modules as needed. If you add a wind turbine, you can downsize the number of solar modules needed

* Another method is to find the rated amperage for a particular module and multiply it by the battery charging voltage (typically 13 volts). Example: Kyocera's 120-watt panel has a Ipmax rating of 7.1 amps x 13 = 92.3 watts. This gives you a slightly more optimistic number.

** You'll get 120 watts from a 120-watt PV module **only** when using an MPPT charge controller during the 2 hours nearest high noon, and **only** when the surface temperature of the module is below 77 degrees (F)...which is hardly ever. Output is typically derated to 60% - 70% for standard charge controllers (75% - 80% with MPPT) to give you a more accurate number.

Battery Sizing Worksheet

		Average	Example	Winter
1.	Input your Adjusted Ave. Watt-Hours per Day *(from the Power Consumption Worksheet)*.	_____	*3,000*	_____
2.	Input the number of days of battery storage you need (the number of cloudy days in a row).*	_____	*4*	_____
3.	Multiply line 2 by line 1.	_____	*12,000*	_____
4.	Input the battery depth of discharge you are comfortable with: 80% discharge (.80) is maximum; 50% discharge (.50) is better.	_____	*.50*	_____
5.	Divide line 3 by line 4.	_____	*24,000*	_____
6.	Derate batteries for low operating temperatures: select a factor next to your lowest operating temperature and enter here: 80°F – 1.00 | 70°F – 1.04 | 60°F – 1.11; 50°F – 1.19 | 40°F – 1.30	_____	*1.11*	_____
7.	To find your total battery capacity, multiply line 5 by line 6.	_____	*26,640*	_____
8.	Calculate the watt-hour capacity of your selected battery: **voltage x amp hour.** *Example: L-16 is 6 volts x 390 amp hours*	_____	*2,340*	_____
9.	To find the number of batteries you need, divide line 7 by line 8. Adjust the number of batteries to fit your system voltage. *Example: 24-volt system requires sets of 4, 6-volt batteries.*	_____	*12*	_____

* If this number is greater than 5, a good backup generator may be more cost effective than extra batteries.

How to Calculate a Wire Run

It's a perennial question for anyone who has ever used a long extension cord to run a hefty power tool: what size wire do I need to safely carry X amps from point A to point B? It is an especially relevant question if you are calculating a wire run to a solar array. Fortunately, it is a fairly easy question to answer, once you are armed with the right formulas. It is simply a matter of knowing which variables to plug in.

For each of the following formulas we will be using four variables to determine a fifth variable which, to make it more exciting, we will pretend we don't know. Below are the variables along with the values we will plug in for the given examples.

Assume a 48-volt system (single-phase) using copper wire for all examples:

- Wire length: **72.44 feet**
- Amperage in wire, maximum: **40**
- Acceptable voltage drop: **2%, or 0.96** volts for a 48-volt system
- Wire area in circular mils: **66,400** for #2 wire (see chart on next page)
- Specific resistivity x the phase constant: 11 x 2 = **22** *

Formula #1: Determining Voltage Drop

(22 x wire length x amps) ÷ wire area in circular mils = voltage drop
(22 x 72.44 x 40) ÷ 66,400 = 0.96 volts

Formula #2: Determining Wire Length

((Voltage drop x wire area in circular mils) ÷ 22) ÷ amps = wire length
((0.96 x 66,400) ÷ 22) ÷ 40 = 72.44 feet

Formula #3: Determining Wire Size

(wire length x amps x 22) ÷ voltage drop = wire area in circular mils
(72.44 x 40 x 22) ÷ 0.96 = 66,400, or #2 copper wire

Formula #4: Determining Maximum Amperage

((Wire area in circular mils x voltage drop) ÷ wire length) ÷ 22 = maximum amperage
((66,400 x 0.96) ÷ 72.44) ÷ 22 = 40

* For 3-phase systems use 11 x 1.732 = 19.05

Thus by knowing any four variables we can always discover the fifth. Just be sure to plug in the right numbers, particularly the voltage drop, which is determined by multiplying the system voltage by the acceptable voltage-loss percentage. For maximum safety and performance the voltage drop should never be more than 2%:

System Voltage x 2% = Voltage Drop
48 volts x 0.02 = 0.96 volts
24 volts x 0.02 = 0.48 volts
60 volts x 0.02 = 1.20 volts

...and so on and so forth. On the following page is a table I prepared using Formula #2. It is designed to give you a close approximation of maximum wire runs for a 48-volt system over a broad range of amperages and wire sizes. Use the above formulas to refine the numbers for your particular system.

Wire Area in Circular mils	
4/0	212,000
3/0	168,000
2/0	133,000
1/0	106,000
#1	83,700
#2	66,400
#3	52,600
#4	41,700
#5	33,100
#6	26,300
#8	16.500

48-Volt System with a 2% Acceptable Voltage Drop

watts	amps	Maximum One-Way Distance (in feet) for Various Copper Wire Sizes						
		#8	#6	#4	#2	1/0	2/0	3/0
1,000	20.83	35	55	87	139	222	278	352
1,500	31.25	23	36	58	93	148	186	235
2,000	41.66	17	27	44	70	111	140	176
2,500	52.08	14	22	35	56	88	112	141
3,000	62.50	--	18	29	46	74	93	118
3,500	72.92	--	16	25	40	63	80	101
4,000	83.33	--	--	22	35	56	70	88

-- Exceeds Ampacity

Microhydro Formulas & Conversion Factors

Gross Head x Flow x System Efficiency x C = Power

Example: 50 ft. x .223 cfs x 0.55 x 0.085 = .52 kW (or 520 watts)

- **Gross head** is the actual distance of drop from the intake to the turbine, not taking into account the friction developed in the penstock (in feet or meters).
- **Flow** is measured in cubic feet per second (cfs), or cubic meters per second (m^3/s).
- **System efficiency** will be between 40% and 70%.
- **C (the constant)** is 0.085 when using cfs; 9.81 when using m^3/s.

Finding the Exact Efficiency of Your System

If you do install an actual system, you can see how close you were to guessing the efficiency factor by running the formula backwards:

True Efficiency Factor = kW ÷ (Gross Head x Flow x C)

Helpful Microhydro Conversion Factors

1 cubic foot (cf) = 7.48 gallons

1 cubic foot per second (cfs) = 448.8 gallons per minute (gpm)

1 cubic foot (cf) = 0.028 cubic meters (m^3)

1 cubic meter (cm) = 35.3 cubic feet (cf)

1 cubic meter per second (m^3/s) = 15,842 gallons per minute (gpm)

1 meter (m) = 3.28 feet

1 foot = 0.3048 meters (m)

1 pound per square inch (psi) = 2.31 feet of head

1 kilowatt (kW) =1.34 horsepower (hp)

1 horsepower (hp) = 746 watts

Resources

The information listed below will give you a good start in the right direction. The internet is an excellent tool for finding for new information and resources in this ever-changing business of renewable energy.

Renewable Energy Manufacturers Mentioned

Bergey Wind Power
www.bergey.com

Bogart Engineering
www.bogartengineering.com

Concorde Battery Corporation
www.concordebattery.com

Controlled Energy Corporation
www.cechot.com

Enertia Building Systems
www.enertia.com

Enphase Energy
www.enphase.com

Fronius USA
www.fronius.com

GeoSystems
www.geosystemsghp.com

Golden Grain Stoves
www.goldengrainstove.com

Grundfos Pumps Corporation
www.grundfos.com

Harris Hydroelectric
www.harrishydro.com

Hawker Batteries
www.hawkerpowersource.com

HS-Tarm
www.woodboilers.com

Kyocera Solar
www.kyocerasolar.com

MK Battery
www.mkbattery.com

OutBack Power Systems
www.outbackpower.com

Proven Energy
www.provenenergy.com

PV Powered
www.pvpowered.com

Sharp Solar Systems
http://solar.sharpusa.com

SMA
www.sma-america.com

SolaHart
www.solahart.com

Southwest Windpower Inc.
www.windenergy.com

Takagi
www.takagi.com

Trojan Battery Company
www.trojan-battery.com

Tulikivi
www.tulikivi.com

Your Solar Home
www.yoursolarhome.com

Xantrex Technology
www.xantrex.com

Education & Classes

Institute for Solar Living
www.solarliving.org

Midwest Renewable Energy Association
www.the-mrea.org

Solar Energy International
www.solarenergy.org

Organizations

American Solar Energy Society
www.ases.org

American Wind Energy Association
www.awea.org

Burn Wise
www.epa.gov/burnwise

Center for Renewable Energy &
Sustainable Technology
www.crest.org

Database of State Incentives for
Renewable Energy
www.dsireusa.org

Efficient Windows Collaborative
www.efficientwindows.org

Energy Star (energy ratings)
www.energystar.gov

Find A Solar Professional
www.findsolar.com

Florida Solar Energy Center
www.fsec.ucf.edu

Geothermal Heat Pump Consortium
www.geoexchange.org

Interstate Renewable Energy Council
www.irecusa.org

Masonry Heater Association
www.mha-net.org

NAHB National Green Building Program
www.NAHBgreen.org

National Renewable Energy Lab
www.nrel.gov

North American Board of Certified
Energy Practitioners (NABCEP)
www.nabcep.org

Renewable Energy World
www.renewableenergyworld.com

Sandia's Photovoltaics Program
www.sandia.gov/pv

SRCC (Solar Rating and Certification
Corporation)
www.solar-rating.org

U.S. Department of Energy's EERE
(Energy Efficiency & Renewable Energy)
www.eere.energy.gov

U.S. Solar Radiation Resource Maps
*http://rredc.nrel.gov/s4olar/old_data/
nsrdb/redbook/atlas*

Wind Energy Maps/Tables
*www.windpoweringamerica.gov/
windmaps/*

Other Renewable Energy books by PixyJack Press

*Got Sun? Go Solar, 2nd Edition: Harness
Nature's Free Energy to Heat and Power
Your Grid-Tied Home*
by Rex A. Ewing and Doug Pratt

*The Smart Guide to Geothermal: How to
Harvest Earth's Free Energy for Heating
and Cooling*
by Don Lloyd

*HYDROGEN – Hot Stuff, Cool Science, 2nd
Edition: Discover the Future of Energy*
by Rex A. Ewing

*Careers in Renewable Energy:
Get A Green Energy Job*
by Gregory McNamee

*Crafting Log Home Solar Style: An
Inspiring Guide to Self-Sufficiency*
by Rex A. and LaVonne Ewing

Glossary

Absorption Stage A stage of the battery-charging process performed by the charge controller, where the batteries are held at the bulk-charging voltage for a specified time period, usually one to two hours.

Air-Source Heat Pump The most common type of heat pump. In heating mode, the heat pump absorbs heat from the outside air and transfers the heat to the inside of building. In the cooling mode the heat pump absorbs heat from inside the building and removes it to the outside air. An air-conditioning unit is a type of air-source heat pump.

Alternating Current (AC) Electric current that reverses its direction of flow at regular intervals, usually many times per second; common household current is AC.

Alternative Energy Energy that is not popularly used and is usually environmentally sound, such as solar or wind energy, hydrogen fuel, or biodiesel. *See also* Renewable Energy.

Amorphous Solar Cell Type of solar cell constructed by using several thin layers of molten silicon. Amorphous solar cells perform better in sub-optimal lighting conditions, but need more surface area than conventional crystalline cells to produce an equal amount of power.

Ampere (Amp) Unit of electrical current, thus the rate of electron flow. One volt across one ohm of resistance is equal to a current flow of one ampere.

Ampere Hour (AH) A current of one ampere flowing for one hour. Used primarily to rate battery capacity and solar or wind output.

Array *See* Photovoltaic Array.

Battery Electrochemical cells enclosed within a single container and electrically inter-connected in a series / parallel arrangement designed to provide a specific DC operating voltage and current level. Batteries for PV systems are commonly 6- or 12-volts, and are used in 12, 24 or 48-volt operations.

Battery Cell The basic functional unit in a storage battery. It consists of one or more positive electrodes or plates, an electrolyte that permits the passage of charged ions, one or more negative electrodes or plates, and the separators between plates of opposite polarity.

Battery Capacity Total amount of electrical current, expressed in ampere-hours (AH), that a battery can deliver to a load under a specific set of conditions.

Battery Life Period during which a battery is capable of operating at or above its specified capacity or efficiency level. A battery's useful life is generally considered to be over when a fully charged cell can only deliver 80% of its rated capacity. Beyond this point, the battery capacity diminishes rapidly. Life may be measured in cycles and/or years, depending on the type of service for which the battery was designed.

Blocking Diode A semiconductor connected in series with a solar module or array, used to prevent the reverse flow of electricity from the battery bank back into the array when there is little or no solar output. Think of it as a one-way valve that allows electrons to flow forward but not backward.

Btu (British Thermal Unit) The energy required to raise the temperature of one pound of water from 39 to 40 degrees F.

Building-Integrated PV (BIPV) Where PV is integrated into a building, replacing conventional materials, such as siding, shingles or roofing panels.

Bulk Stage Initial stage of battery charging, where the charge controller allows maximum charging in order to reach the bulk voltage setting.

Cell *See* Photovoltaic Cell.

Cell Efficiency Percentage of electrical energy that a solar cell produces (under optimal conditions) divided by the total amount of solar energy falling on the cell.

Typical efficiency for commercial cells is in the range of 12 to 15%.

Charge Controller Component located in the circuit between the solar array or wind turbine and the battery bank. Its job is to bring the batteries to an optimal state of charge without overcharging them. Most charge controllers have digital displays to help monitor system status and performance. MPPT charge controllers go a step further by converting excess array voltage into usable amperage.

Circuit A system of conductors connected together for the purpose of carrying an electric current from a generating source, through the devices that use the electricity (the loads), and back to the source.

Circuit Breaker Safety device that shuts off power (i.e. it creates an open circuit) when it senses too much current.

Conductor A material—usually a metal, such as copper—that facilitates the flow of electrons.

Conversion Efficiency See Cell Efficiency.

Current Flow of electricity between two points. Measured in amps.

Depth of Discharge (DOD) The ampere-hours removed from a fully charged battery, expressed as a percentage of rated capacity. For example, the removal of 25 ampere-hours from a fully charged 100 ampere-hour rated battery results in a 25% depth of discharge. For optimum health in most batteries, the DOD should never exceed 50%.

Direct Current (DC) Electrical current that flows in only one direction. It is the type of current produced by solar cells and the only current that can be stored in a battery.

Distributed System A system installed near where the electricity is used, as opposed to a central system—such as a coal or nuclear power plant—that supplies electricity to the electrical grid. A grid-tied residential solar system is a distributed system.

Electrical Grid A large distribution network—including towers, poles, and transmission lines—that delivers electricity over a wide area.

Electric Circuit See Circuit.

Electric Current See Current.

Electricity In a practical sense, the controlled flow of electrons through a conductor. In a scientific sense, the non-gravitational and non-nuclear repulsive and attractive forces governing much of the behavior of charged subatomic particles.

Electrode A conductor used to lead current into or out of a nonmetallic part of a circuit, such as a battery's positive and negative electrodes.

Electrolyte Fluid used in batteries as the transport medium for positively and negatively charged ions. In lead-acid batteries this is a somewhat diluted sulfuric acid.

Electron Negatively charged particle. An electrical current is a stream of electrons moving through an electrical conductor.

Energy The capacity for performing work. A ball resting on the top of a hill is said to have potential energy, while the same ball rolling down the hill is imbued with kinetic energy. Solar cells convert electromagnetic energy (light) from the sun into electrical energy, while wind turbines convert the kinetic energy of the air into first mechanical energy, and then electrical energy.

Energy Audit An inspection process that determines how much energy you use in your home, usually accompanied by specific suggestions for saving energy.

Equalization A controlled process of over-charging non-sealed lead-acid batteries, intended to clean lead sulfates from the battery's plates and restore all cells to an equal state-of-charge.

Evacuated Tube Collector A type of solar hot-water collector that uses "evacuated," or vacuum-sealed, glass tubes to isolate copper channels through which solar-heated water flows.

Flat Plate Collector In solar hot-water systems, the primary unit for collecting the sun's energy. Flat-plate collectors consist of a continuous loop of black copper pipe partially embedded in an insulating material within the frame, and covered with tempered glass.

Float Stage A battery-charging operation performed by the charge controller in which enough energy is supplied to meet all loads, plus internal component losses, thus always keeping the battery up to full power and ready for service. Float voltage is somewhat lower than bulk voltage.

Fossil Fuels Carbon- and hydrogen-laden fuels formed underground from the remains of long-dead plants and animals. Crude oil, natural gas and coal are fossil fuels.

Full Sun Scientific definition of solar power density received at the surface of the earth at noon on a clear day. Defined as 1,000 watts per square meter (W/m^2). Reality varies from 600 to 1,200 W/m^2, depending on latitude, altitude, and atmospheric purity.

Geothermal Literally, heat from the Earth. This includes the seasonal solar heat stored in the top several feet of the Earth's crust, as well as the heat that naturally conducts toward the surface from the Earth's mantle and core.

Geothermal Heat Pump (GPH) *See* Ground-Source Heat Pump.

Greenhouse Effect A warming effect that occurs when heat from the sun becomes trapped in the Earth's atmosphere due to the heat-absorbing properties of certain (greenhouse) gases.

Greenhouse Gases Gases responsible for trapping heat from the sun within the Earth's atmosphere. Water vapor and carbon dioxide are the most prevalent, but methane, ozone, chlorofluorocarbons and nitrogen oxides are also important greenhouse gases.

Grid *See* Electrical Grid.

Grid-Connected PV System A solar PV system that is tied into the utility's electrical grid. When generating more power than necessary to power all its loads, the system sends the surplus to the grid. At night the system draws power from the grid.

Ground-Source Heat Pump A type of heat pump that uses the natural heat storage ability of the earth or the groundwater to heat or cool a building. Also called geothermal heat pump (GPH).

Heat Exchanger A device used to transfer heat from one reservoir of fluid to another.

Heat Pump *See* Ground-Source Heat Pump or Air-Source Heat Pump.

Hertz (Hz) A unit denoting the frequency of an electromagnetic wave, equal to one cycle per second. In alternating current, the frequency at which the current switches direction. In the U.S. this is usually 60 cycles per second (60 Hz).

Hybrid System Power-generating system consisting of two or more subsystems, such as a wind turbine or diesel generator, and a photovoltaic system.

Hydronic A heating system that uses circulating hot water to transfer heat from a boiler, water heater, or solar collector to the inside of a building.

Insolation Measure of the amount of solar radiation reaching the surface of the Earth. According to NREL, "this term has been generally replaced by solar irradiance because of the confusion of the word with insulation." *See* Irradiance.

Inverter Component that transforms the direct current (DC) flowing from a solar system or battery to alternating current (AC) for use in the home. Also called a power inverter.

Irradiance Rate at which radiant energy arrives at a specific area of the Earth's surface during a specific time interval. Measured in W/m^2.

I-V Curve A graph that plots the current versus the voltage from a solar cell, as the electrical load (or resistance) is increased from short circuit (no load) to open circuit (maximum voltage). The shape of the curve characterizes the cell's performance. Three important points on the I-V curve are the open-circuit voltage, short-circuit current, and peak or maximum power (operating) point.

Junction Box (J-Box) Enclosure on the back of a solar module where it is connected (wired) to other solar modules.

Kilowatt (kW) Unit of electrical power, equal to 1,000 watts.

Kilowatt-Hour (kWh) 1,000 watts being used over a period of one hour. The kWh is the usual billing unit of energy for utility companies.

Life-Cycle Cost Estimated cost of owning, operating, and disposing of a system over its useful life.

Load Anything that draws power from an electrical circuit.

Maximum Power Point Tracking (MPPT) Technology used by direct grid-tied inverters and many charge controllers to convert, through the use of DC-DC power converters, excess array voltage into usable amperage by tracking the optimal power point of the I-V curve.

Megawatt (MW) One million watts, or 1,000 kilowatts. Commercial power plants and wind farms are usually rated in megawatts.

Microhydro A home-scale hydroelectric system designed to produce electricity by rerouting a portion of a stream's flow through a turbine. The DC power the turbine produces is stored in batteries in the same way solar or wind energy is stored.

Module *See* photovoltaic module.

Monocrystalline Solar Cell Type of solar cell made from a thin slice of a single large silicon crystal. Also known as single-crystal solar cell.

Multicrystalline Solar Cell *See* polycrystalline solar cell.

NABCEP (North American Board of Certified Energy Practitioners) A volunteer board of representatives from the renewable energy industry dedicated to developing and implementing quality credentialing and certification programs for renewable-energy practitioners, particularly installers of PV, wind and solar hot-water systems.

National Electrical Code (NEC) The U.S. minimum inspection requirements for all types of electrical installations, including solar/wind systems.

NEMA (National Electrical Manufacturers Association) The U.S. trade organization which sets standards for the electrical manufacturing industry.

NREL (National Renewable Energy Laboratory) Based in Golden, Colorado, NREL is the principal research laboratory for the DOE Office of Energy Efficiency and Renewable Energy. Operated by Midwest Research Institute and Battelle, NREL concentrates on studying, testing and developing renewable energy technologies.

Net Metering A practice used in conjunction with a solar- or wind-electric system. The electric utility's meter tracks the home's net power usage, spinning forward when electricity is drawn from the utility, and spinning backward when the solar or wind system is generating more electricity than is currently needed to run the home's loads.

Ohm Measure of the resistance to current flow in electrical circuits, equal to the amount of resistance overcome by one volt producing a current of one ampere.

Orientation Term used to describe the direction that a solar module or array faces. The two components of orientation are the tilt angle (the number of degrees the panel is raised from the horizontal position) and the aspect angle, (the degree by which the panel deviates from facing due south).

Parallel Connection Wiring configuration whereby the current is given more than one path to follow, thus amperage is increased while voltage remains unchanged. In DC systems, parallel wiring is positive to positive (+ to +) and negative to negative (- to -). *See also* Series Connection.

Passive Solar Home Home designed to use sunlight for direct heating and lighting, without circulating pumps or energy conversion systems. This is achieved through the use of energy-efficient materials (such as windows, skylights and Trombe walls) and proper design and orientation of the home.

Peak Load Maximum amount of electricity being used at any one point during the day.

Pelton Wheel A type of water wheel used to turn a microhydro turbine. A Pelton wheel uses slightly offset cups to catch water and transfer its kinetic energy to the shaft of the turbine.

Penstock In microhydro systems, a pipe that conveys water under pressure from the water source to the microhydro turbine. Commonly made of PVC pipe.

PEX Tubing Cross-linked polyethylene. Type of tubing used to circulate water or glycol through a floor's hydronic heating system. Also used in geothermal heat pump systems.

Photon Basic unit of light. A photon can act as either a particle or a wave, depending on how it's activity is measured. The shorter the wavelength of a stream of photons, the more energy it possesses. This is why ultraviolet (UV) light is so destructive, while infrared (IR) generally is not.

Photovoltaic (PV) Refers to the technology of converting sunlight directly into electricity through the use of photovoltaic (solar) cells.

Photovoltaic Array A system of interconnected PV modules (solar panels) acting together to produce a single electrical output.

Photovoltaic Cell The basic unit of a PV (solar) module. Crystalline photovoltaic cells produce an electrical potential of around 0.5 volts. The higher voltages typical in PV modules are achieved by connecting solar cells together in series.

Photovoltaic Module Collection of solar cells joined as a unit within a single frame. Commonly called a "solar panel."

Photovoltaic System Complete set of interconnected components (including a solar array, inverter, etc.) designed to convert sunlight into usable electricity.

Polycrystalline Solar Cell Type of solar cell made from many small silicon crystals (crystallites). Because of the numerous grain boundaries, devices that employ this design will operate with slightly reduced efficiency. Also known as a multi-crystalline solar cell.

Primary Combustion The initial stage of burning biomass. Primary combustion begins at around 550 degrees F and converts about half of the energy of the biomass into heat.

PV Photovoltaic.

Rated Power Nominal power output of inverters, solar modules or wind turbine; some units cannot produce rated power continuously.

Renewable Energy (RE) Energy obtained from sources that are essentially inexhaustible (unlike, for example, fossil fuels, of which there is a finite supply). Renewable sources of energy include conventional hydroelectric power, wood, waste, geothermal, wind, photovoltaic, and solar-thermal energy.

Secondary Combustion The burning of combustible gases that are not burned during primary combustion. Secondary combustion takes place at higher temperatures than primary combustion, generally around 1,100 degrees F. *See also* Wood Gas.

Semiconductor Material that has an electrical conductivity in between that of a metal and an insulator. Typical semiconductors for PV cells include silicon, gallium arsenide, copper indium diselenide, and cadmium telluride.

Series Connection A wiring configuration where the current is given but one path to follow, thus increasing voltage without changing the amperage. Series wiring is positive to negative (+ to -) or negative to positive (- to +). *See also* Parallel Connection.

Silicon (Si) The most common semiconductor material used in the manufacture of PV cells. In the periodic table, it is element number 14, positioned between aluminum and phosphorus.

Single-Crystal Silicon *See* Monocrystalline Solar Cell.

Solar Cell *See* Photovoltaic Cell.

Solar Collector *See* Flat Plate Collector or Evacuated Tube Collector.

Solar Energy Energy from the sun. Virtually all energy on Earth (including solar, wind, hydroelectric and even fossil-fuel energy) originated as solar energy.

Solar Insolation *See* Irradiance.

Solar Module *See* Photovoltaic Module.

Solar Panel Common term used to describe a PV (solar) module. "Solar panel" refers to both photovoltaic modules, used for making electricity, and solar hot-water panels, used to augment a home's heating system. (Compare with flat-plate collector.)

Solar Power *See* Solar Energy.

SRCC (Solar Rating & Certification Corporation) A non-profit organization whose primary purpose is to provide authoritative performance ratings, certifications and standards for solar thermal products.

Stand-Alone System A solar-electric system that operates without connection to the utility grid, or another supply of electricity. Typically, unused daylight energy production is stored in a battery bank to provide power at night. Stand-alone systems are used primarily in remote locations, such as mountain areas, ocean platforms or communication towers.

Thin Film *See* Amorphous Solar Cell.

Tilt Angle The angle of inclination of a module measured from the horizontal. The most productive tilt angle is one in which the surface of the module is exactly perpendicular to the sun's rays.

Volt (V) A unit of electrical force, analogous to the water pressure within a garden hose. It is equal to the amount of electromotive force that will cause a steady current of one ampere to flow through a resistance of one ohm.

Watt (W) Unit of electrical power used to indicate the rate of energy produced or consumed by an electrical device. One ampere of current flowing at a potential of one volt produces one watt of power. Wind turbines and PV modules are often rated in watts.

Watt-hour (Wh) Unit of energy equal to one watt of power being used or produced for one hour.

Wind Energy The kinetic energy present in wind, measured in watts per square meter (W/m^2). Wind turbines convert the kinetic energy into mechanical energy through the use of propeller blades, which in turn drive an alternator to produce electricity.

Wood Boiler Also called wood-fired hydronic furnace. A device, generally located outside the house or building, that the burns wood to heat water for use in forced air or hydronic radiant heating systems.

Wood Gas Gases not burned during the primary combustion of wood, consisting mainly of hydrogen (H_2), carbon monoxide (CO) and methane(CH_4). When free of contaminants such as tar, wood gas is a suitable fuel for internal combustion engines.

Wood Gas Generator A device that converts wood into a mixture of secondary-combustion gases that can then be burned in an internal combustion engine.

Index

About the Author

From his solar- and wind-powered studio in the Colorado Rockies, Rex A. Ewing has penned a number of best-selling books on renewable energy, including *Power With Nature*; *HYDROGEN—Hot Stuff, Cool Science*; and *Got Sun? Go Solar*. For aspiring log home owners who want to incorporate solar energy into their dream home, he teamed up with his wife to produce *Crafting Log Homes Solar Style: An Inspiring Guide to Self-Sufficiency*. His magazine articles can be found in log home and sustainable living magazines.

Before moving to the mountains to handcraft a log home and concentrate on his writing, Ewing raised grass hay and high-strung Thoroughbred race horses in the Platte River valley. Whenever his employees were persistent enough to corral him behind a desk, he served as CEO of a well-respected equine nutrition firm, where he formulated and marketed a successful line of equine supplements worldwide. In 1997 he wrote a best-selling book on horse nutrition: *Beyond the Hay Days: Refreshingly Simple Horse Nutrition*; the expanded 2nd edition was released in 2004.

When he's not writing, he and his wife LaVonne can be found trekking through the backcountry, canoeing with their dogs, or enjoying the 50-mile view from their deck.

www.RexEwing.com

100% solar & wind powered since 1999

PixyJack Press INC

Wholesale Orders Welcome
PO Box 149, Masonville, CO 80541 USA
www.PixyJackPress.com   *info@pixyjackpress.com*